BIOCHEMICAL
ECOLOGY OF
WATER POLLUTION

BIOCHEMICAL ECOLOGY OF WATER POLLUTION

Patrick R. Dugan

Professor and Chairman
Department of Microbiology
The Ohio State University
Columbus, Ohio

ℂℙ PLENUM PRESS • NEW YORK–LONDON • 1972

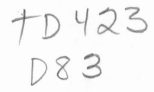

First Printing – January 1972
Second Printing – January 1973

Library of Congress Catalog Card Number 72-167676
ISBN 0-306-30540-2

© 1972 Plenum Press, New York
A Division of Plenum Publishing Corporation
227 West 17th Street, New York, N.Y. 10011

United Kingdom edition published by Plenum Press, London
A Division of Plenum Publishing Company, Ltd.
Davis House (4th Floor), 8 Scrubs Lane, Harlesden, NW10 6SE, London, England

PREFACE

Biochemical ecology is here presented only in the context of water pollution. This is not to minimize the importance of land animals and plants in their environment or the significance of air pollution as it relates to ecology. It merely indicates that water pollution is a problem of sufficiently broad magnitude to warrant consideration by itself.

Water pollution is a problem which requires the attention of a variety of disciplines. The presentation tends therefore to follow the problem approach, as do most interdisciplinary topics. An appreciation of various viewpoints is needed among chemists, ecologists, economists, engineers, lawyers, limnologists, managers, microbiologists, and politicians, whose communications are often "hung up" in each other's jargon.

Perhaps the presentation is too elementary at times. This was done in an attempt to bridge the diverse backgrounds of those concerned with the subject. It is hoped that engineers, economists, biologists, public servants, and others will gain a greater appreciation of the interrelationship of gross observations and biological events that occur at the cellular and molecular level. Lack of such understanding is, to a large extent, the reason for our present environmental condition. At other times the presentation is perhaps too technical. This was done on the assumption that some information on chemical details may not be readily available but is desirable for an "in-depth" appreciation of the biochemical events encountered in water pollution.

The pattern of presentation is to give background information in relatively simple terms and then to support it with more detailed data. In this approach I would argue that it is the significance of reactions in the aquatic environment that is of importance. Consequently, the activities of organisms have significance to water pollution, whereas the numbers and names of the organisms are relegated to secondary considerations.

When specific reactions are discussed, it is implied that they are likely to proceed in the aquatic environment. For example, *Pseudomonas* species are very common to both soil and water. Their metabolic activities would likely be encountered in water, although they have been studied in the laboratory. I have avoided, wherever possible, using biochemical information that would not likely be encountered in the aquatic ecosystem.

In addition to the audience already mentioned, it is hoped that this monograph will be of value to both undergraduate and graduate students with an interest in the aquatic environment and to those individuals who avow an interest in the social-political-economic ramifications of an unbalanced ecosystem.

I would like to acknowledge my wife's patient assistance in the preparation of the manuscript, and to thank Dr. Jorgen Birkeland for valuable criticism and suggestions and Dr. Chester Randles for aid in preparing Chapter 11.

PATRICK R. DUGAN
September, 1971

CONTENTS

Part II

BIOCHEMICAL CONSIDERATIONS

Part III

MAJOR ECOLOGICAL PROBLEMS

PART I

THE WATER POLLUTION PROBLEM

Chapter 1

SIGNIFICANCE OF POLLUTION

Water pollution is significant only when it influences living or biological systems either directly or indirectly. In a broad sense, it can be depicted as a normal consequence of the growth of organisms including man in or near the aquatic habitat. The unique physical and chemical properties of water have allowed life to evolve in it. The following quote from Szent-Gyorgyi (1958) illustrates this point of view: "That water functions in a variety of ways within a cell can not be disputed. Life originated in water, is thriving in water, water being its solvent and medium. It is the matrix of life." All biological reactions occur in water, and it is the integrated system of biological metabolic reactions in an aqueous solution that is essential for maintenance of life. This also is true of air pollutants in that their impact on organisms occurs only when the pollutants are placed in solution (e.g., in the human lung).

Water is the most preponderant chemical found within any freely metabolizing cell, and in bacterial cells the content ranges from 75 to 90 percent, with an average of about 80 percent. In addition to the importance of water in contributing to total cell mass, it is the most versatile of all chemicals in the cell by virtue of its participation in regulatory mechanisms (e.g., osmotic phenomenon and salt balance) and intermediary metabolism (e.g., hydrolytic reactions) and as a structural component (e.g., maintaining turgidity, rigidity, and tertiary structure of macromolecules).

Whether we view life at the cellular, organismic, or population levels, for our purposes all levels can be equated to the biomass of living material. In biochemical terms, the biomass requires both an energy supply to be used as fuel for carrying out the various activities of cells, organisms, or populations, and structural components for the purpose of assembling new cells, organisms, or populations. When living systems expend an energy-rich raw material for growth and activity, it is axiomatic that a lower-energy by-

3

product or waste material will result from the process, and no species of organism can live on its own waste products. This by-product excreta or waste material constitutes a pollutant when it enters the air or water. It is therefore inevitable that living things produce pollutants and generally in proportion to the size, rate of activity, and efficiency of the biomass. It is implicit that virtually all biological and biochemical reactions are catalyzed by enzymes, and to interfere with enzymes is to interfere with life processes, although enzyme inactivation is not the only way in which pollutants interfere with or influence life processes.

Again, it is the uniqueness of water properties that in a broad sense allows pollutants to accumulate in water. To quote Revelle (1969), "the fluid character of water means that the oceans (and lakes) fill all the low places of the earth. Because of this geographical fact, the oceans (and lakes) are ultimate receptacles of the wastes of the land; including wastes that are produced in ever increasing amounts by human beings and their industries."

Although pollution is produced by the activities of organisms, it is usually recognized only when it adversely affects other living organisms; e.g., fish are killed, algal growth is enhanced to bloom proportions which then insult human aesthetic values, people contract a disease. Generally, the reference point for identifying and assessing pollution is the impact it has on human interests. However, we have become sophisticated enough to realize that anything which indirectly influences man's well-being is as important as man himself and that pollution which ultimately influences man's well-being, although indirectly, is also a matter of survival. It is the myriad of diverse organisms acting in concert which allows the continuing recycling of the finite amount of each chemical element available on earth. When one or another of the elements is prohibited from recycling by the elimination of species of organisms responsible for a particular biochemical link in the system, it will accumulate in a particular chemical form and put the ecosystem out of balance. Herein lies the ecological significance of the pollution problem. The sum total is that pollution puts the human environment out of balance, and the scheme of nature is such that it will react to re-establish a balance. Human activities which put the ecosystem out of balance and endanger species, etc., represent part of a normal continuously changing ecosystem and therefore a natural consequence of human life. Regardless of the philosophical point of view, we can predict that newly established balances will be markedly different from life and surroundings as we know them today, if rampant polluting is allowed to continue at its present pace.

There are many intermediary stages in which various groups of organisms act upon the by-products or wastes of other organisms. What is one organism's waste or poison is another's food. Genetic, nutritional, and metabolic processes function as a natural selection process and allow shifts of populations in response to alterations in the environment and indeed exert an altering effect on the environment. Individual ecosystem components, as we know them, depend on or require waste products (pollutants) from another segment of the ecosystem. This is the way in which living systems evolve and maintain the balance. Therefore, once the nutritional energy balance relationships have formed, these segments of the ecosystem are dependent upon the pollutants from others. If the flow of pollutants at this level is stopped, this segment of a dynamic ecosystem would come to equilibrium and cease to exist. Complicated interactions in the environment are difficult to predict or evaluate and have provided a stimulus for enlisting the techniques of systems analysis.

It is the ability of certain organisms to utilize the nutritionally rich polluted environments, such as domestic sewage, that is exploited by man in biological waste treatment processes. In this regard, microbes are well suited for exploitation because of their high metabolic rates per unit cell mass. However, microbial activity, acting on sewage as nutrients for further growth, would proceed in a stream or lake if the microbes were not confined to the waste treatment plant by engineering design.

Specific chemicals, other than general poisons, are not readily definable as pollutants. Rather, it is the quantity or concentration of specific compounds in an isolated situation that must be considered in relation to an observed effect. For example, phosphate is an absolute necessity for all life, yet it is of primary concern today as a water pollutant when it is a growth-limiting nutritional factor, because the excessive amounts which have been deposited in the environment promote excessive growth of algal cells in our lakes. Pollutants then are akin to our concept of weeds—they are chemicals out of place. It would seem to follow that pollution as a state of the environment must be discussed in specific terms related to individual circumstances and, contrary to popular opinion, should not be over-generalized. However, water pollutants for the most part are chemicals dissolved or suspended in water which elicit an environmental response that is objectionable. Some pollutants are physical factors and not chemicals; e.g., heat and radiation are physical factors and exert a marked effect on bio-chemical reactions.

Part of our difficulty in assessing the significance of pollution is due to our lack of present knowledge of limits of tolerance. There has been a tendency to make judgments on the basis of acute toxicity values such as fish kills. It is now evident that accumulative threshold levels of combinations of pollutants must be considered as well as sublethal chronic effects. For example, it is known that air pollutants influence susceptibility of humans to upper respiratory virus infection, age has a direct bearing on nutrition of individuals, and prior exposure to detergents has a decided effect on the susceptibility of fish to chlorinated pesticides.

Another mistake we have made is in overestimating the carrying capacity of water and air for pollutants. Water acts as a solvent for pollutants, which is the reason it is used in many industrial processes as well as for disposal of domestic wastes. For these reasons, it is important to consider in greater detail the solvent properties of water in addition to its ability to mechanically carry suspended particles from agricultural fields, etc.

Finally, I would argue that the physical and chemical properties of water have a direct bearing on pollution and its effects. One such situation involves water density–temperature relationships, which control turnover of deep lakes. This has far-reaching implications in lake eutrophication because of the role of bottom particulates as surfaces for increased biochemical activity and also because of redistribution of nutrients.

Chapter 2

POLLUTIONAL CONCERNS, CAUSES, AND CONCEPTS

2.1. DISEASE PRODUCTION

There have been five or six general areas of concern over the years which have preoccupied individuals concerned with water contamination and its ecological ramifications.

One was the transmission of disease via the water route as the result of contamination by pathogenic bacteria and protozoans originating in the human intestinal tract. Epidemics of typhoid, dysentery, and other gastrointestinal diseases from "sewage" contamination resulted in the emergence of sanitary microbiology as a discipline in the late 1800s. Sanitary engineering is a technology devoted to doing something about the problem, and its development paralleled that of the science of sanitary microbiology. In recent years, we have come to realize that poliomyelitis and infectious hepatitis as well as several types of intestinal, respiratory, and other virus-caused diseases are also transmitted via the water route.

2.2. ORGANIC POLLUTANTS

2.2.1. Oxygen Consumption

Once we learned to control water-borne disease transmission by modern sanitary practices, particularly through the use of chlorine as an antibacterial and antiviral agent, it became apparent that there was more to be concerned about in sewage. Domestic sewage consists primarily of organic excreta which can be utilized as nutrients by other organisms, particularly microorganisms in the environment. These organisms metabolize the organic components of sewage via oxidation reactions and consume oxygen dissolved in the water during the process. Because oxygen has a relatively low solubility

7

in water, it is rapidly consumed and depleted during waste organic oxidation and the water becomes anaerobic. Once the dissolved oxygen is gone, it is not available for fish and other aerobic organisms, which then die of oxygen deficiencies. It should be pointed out that when several types of aerobic organisms such as fish and bacteria are competing in an ecosystem for oxygen and nutrients, the bacteria and other single-celled microbes are the most competitive. This is due in part to the high surface-to-volume ratio of microorganisms, and, as will be pointed out subsequently in greater detail, the oxidative enzymes are located in the surface membranes of microorganisms.

Oxygen balance in water has been and still is an area of preoccupation. Again, sanitary engineering technology has striven to cope with this problem by designing waste treatment systems, many of which utilize forced aeration in the process. Secondary waste treatment has historically been aimed at removal of oxidizable organic material, usually by employing aerobic microbiological processes (e.g., activated sludge or trickling filters) for removal of biochemical oxygen demand (BOD), and the resultant sludge is often further treated in an anaerobic digester. Success of the aerobic microbiological process depends upon conversion of dissolved organics to microbial cell mass of a type which flocculates and thereby settles. The process coincidentally also removes small particulate material from suspension. Anaerobic treatment is often used to reduce sludge volume via biological dewatering reactions, thereby making the total cell mass more amenable to manipulation and ultimate disposal.

2.2.2. Organic Nutrients

In addition to oxygen consumption, most organic compounds serve as nutrients for microorganisms and more highly evolved life forms in the aquatic habitat. In recent years, the excessive algal growth in water that receives treated sewage effluents has focused attention on the necessity for removal of algae-stimulating nutrients during waste treatment. Consideration has been given to extending treatment or instituting tertiary treatment processes for the purpose of removing the algal growth stimulants that remain after secondary treatment. This includes both organic and inorganic mineral nutrients.

Liebig's law of limiting growth factors (Odum, 1959) states that the rate of growth of an organism (or any biological reaction) is controlled by the factor (nutrient) which is limiting. If the concentration of a nutrient is increased until it is no longer limiting, a different factor becomes limiting, etc. It is then apparent that any possible permutation or combination of nutrient factors may be involved in eliciting algal growth in receiving waters, depending upon other environmental influences and the variety of algal species present.

Although the algae of pollutional significance are photosynthetic autotrophic microorganisms and are capable of growth at the expense of mineral nutrients, carbon dioxide, and light in essentially the same manner as with higher plant forms, they are also capable of utilizing organic compounds as nutrients. Many unicellular green algae (e.g., *Chlorella*) can substitute organic compounds such as glucose for light as their source of energy for growth. Blue-green algae cannot substitute organic compounds as an exclusive source of energy, but they can utilize many organics for cell synthesis in the presence of light. This may be an academic point that is irrelevant in naturally polluted water because light would be available to the cells. The significance of the B vitamins (B_{12}, thiamine, and biotin) as algal growth stimulants has been described by Provasoli (1969). B vitamins are synthesized by many species of bacteria and are present in any organic waste effluent.

2.3. MINERAL POLLUTANTS

The third general area of interest is with mineral pollution. Much of this type of pollution is related to our technological development. Several different aspects can be considered as part of this problem.

2.3.1. Mineral Nutrients

As indicated under the topic of organic pollutants, organic and inorganic molecules which stimulate biological responses are not separate ecological entities. All cells need relatively high concentrations of nitrogen, phosphate, potassium, magnesium, sulfur, calcium, and iron, in addition to traces of manganese, zinc, copper, and sodium. Plants including the algae also require boron, chloride, and molybdenum ions. They usually utilize either nitrate or ammonium ions as the nitrogen source, although organic nitrogen such as urea or amino acids will also satisfy the nitrogen requirement in most cases. The source of sulfur most commonly utilized by plant cells is the mineral sulfate, although some sulfur can be derived from sulfur-containing amino acids or sulfide.

One source of minerals is to some extent a consequence of our waste treatment technology. Some of the biochemical reactions involved in organic oxidations can be summarized according to the following summary reactions:

$$C_6H_{12}O_6 \text{ (sugar)} + 6\,O_2 \rightarrow 6\,CO_2 + 6\,H_2O \text{ (complete oxidation)} \quad (2.1)$$

$$\text{tryptophan (amino acid)} + 5\,O \rightarrow \text{indole} + \text{pyruvate} + NH_3$$
$$\text{(incomplete oxidation)} \quad (2.2)$$

$$H_2N-\overset{\overset{\displaystyle O}{\|}}{C}-NH_2 \text{ (urea)} + H_2O \rightarrow CO_2 + 2\,NH_3 \text{ (hydrolytic)} \quad (2.3)$$

The point is that mineral compounds such as CO_2, NH_3, PO_4^{-3}, and SO_4^{-2} are all products of the oxidation of organic materials and all serve as nutrients for subsequent growth of other organisms. These products are also added directly to the environment as crop fertilizers. When they find their way into receiving waters via erosion and other mechanisms, they act as fertilizers for the photosynthetic microorganisms such as green and blue-green algae, diatoms, and also larger plants. Aquatic photosynthetic organisms can also assimilate large quantities of certain organic pollutants which may not have been oxidized. This complicates the problem of algal growth stimulation in our lakes, but in general the problem has dictated an interest in mineral pollution which is related to the more complex consideration of accelerated eutrophication of lakes.

Another major source of mineral nutrients results from surface water runoff. Both urban drainage and agricultural drainage contribute to nutritional enrichment of receiving waters. Surface runoff occurs when precipitation exceeds the rate of absorption by soil.

VanDenburgh and Feth (1965) reported that 11 western river basins contained an average annual solute load of 58 tons per square mile of river basin as the result of erosion. Biggar and Corey (1969) point out that the highest quantities of eroded solutes occur in areas of abundant precipitation and runoff, whereas concentrations of solute are highest in areas of low precipitation, and that the greatest suspended sediment loads are found in areas of intermediate effective precipitation. The latter investigators discuss the movement of nutrients in soluble and particulate forms. Dugan *et al.* (1970*a*,*b*) have described the significance of some of the interactions among suspended microparticulates—dissolved chemicals—and microorganisms as they relate to pollutional effects in a lake.

Urban runoff is also a major contributor of mineral and other pollutants to surface water. This problem is complicated in that 18 percent of the United States communities having sewer systems (1920) combine storm runoff with their domestic sewage (Weibel, 1969). The Federal Water Pollution Control Administration (1969) (the name of which was recently changed from FWPCA to Environmental Protection Agency, EPA) indicates that a total of 1329 municipal jurisdictions serving over 54 million people are wholly or partially served by combined sewers as compared to an estimated 60 to 65 million people in the United States who are served by separate storm sewers. This suggests that pollutants flushed from the streets enter the sewage system and add to the burden of the waste treatment plant. During periods of high storm runoff, the flow of water through a combined system quite often exceeds the flow capacity of the treatment plant and the combined sewage must be diverted around the facility. In this event, all pollutants including raw sewage flow directly into receiving lakes and streams without any treatment. It is also easier to treat a more concentrated sewage than a diluted sewage because it has a higher nutrient value and a lower volume to handle.

An estimate of the extent of pollution by runoff has been made by the FWPCA (1969) during a study of 18 test areas of representative occupancy. Some of the findings show an average street litter load of 4.7 lb per 100 ft of curb per day in commercial areas and 2.4 lb per 100 ft per day in single-family residential areas. The most significant water pollutional component was dust and dirt (1.5 lb/100 ft/day) of which 3 percent was soluble. The soluble fraction contained an average of 5 mg BOD per gram, 40 mg chemical oxygen demand (COD) per gram, 0.48 mg nitrogen forms per gram, and less than 0.05 mg phosphate per gram. This has been interpreted, on the basis of a city the size of Chicago, to have the BOD equivalent of 1 percent of the raw sewage pollution loading or 5 percent of the secondary treatment effluent loading. Assuming that a 14-day accumulation of material occurred prior to the next 2-hr storm, the peak flush effect on receiving water would be equivalent to 160 percent of the raw sewage BOD load and 800 percent of the secondary treated load.

Ice control salt resulted in an expressway runoff that averaged 1.4 percent chloride and had a maximum of 2.5 percent chloride during snow melt runoff. Other chemicals such as pesticides and fertilizers are known to be present in urban runoff, but quantitative data are not available for large sample areas.

Atmospheric dust fall is reported to average as much as 300,000 lb per square mile per year in a city. Air pollutants from motor vehicles, industry, and power plants (e.g., CO, SO_2, NO, hydrocarbons, and particulate matter) eventually either settle to the earth surface or are scavenged by snow and rain. The washed-out gases and settled aerosols (smoke) add to water pollution either directly or via storm water runoff.

The trend toward separation of storm runoff from sewage systems is prevalent in suburban developments. However, the attitude that runoff is "clean water" and therefore can be channeled from catch basins directly into receiving streams may be quite presumptuous in view of the enormous amount of refuse which is flushed into runoff. The suburban practice of placing grass cuttings in catch basins as well as the practice of curbing dogs and other pets will result in a shift toward organic pollution in urban or suburban runoff. The result will be a network of open sewers, and these could become as objectionable to people as an open domestic sewer. Grass cuttings and other seemingly innocuous debris will allow production of odors, bacterial slime, and aesthetic insults when acted on biochemically, as would any other decaying vegetable matter. It may be prudent to reconsider the tradeoffs involved in separation of storm runoff from sewage systems. In this regard, we are fortunate that decomposition of vegetable tissue produces less objectionable products than putrefaction of animal tissue.

2.3.2. Acid Mine Drainage

Acid mine drainage is another mineral pollution problem of major consequence which encompasses an area of approximately 10^5 square miles across the 11 states that make up the Appalachia coal mining region as well as other isolated mining areas of the United States and around the world. The most problem-causing pollutants in acidic mine drainage are iron, sulfate, and hydrogen ions (acidity) which are produced as the result of oxidation of pyritic minerals left behind after coal, chalcopyrite, and other marketable ores have been removed. Other mineral ions are also present in high concentrations in mine water, but these have not been considered to be particularly potent pollutants. Iron pyrite (FeS_2) is a mineral composed of reduced iron (Fe^{+2}) and reduced sulfur (S^{-2}). It is usually found in proximity to coal seams and is left exposed to atmospheric moisture and oxygen either on the surface or in the mines after the coal has been removed. Large piles of refuse pyrite and low-grade coal which accumulate on the land surface are referred to as "gob piles" in mining jargon. A single gob pile often encompasses 50 to 100 acres. Acidophilic bacteria in the *Thiobacillus–Ferrobacillus* group oxidize pyrite as a nutrient and derive energy from the process according to the following schematic reactions (2.4, 2.5, 2.6):

$$Fe^{2+} \rightarrow Fe^{3+} + \text{electron} \tag{2.4}$$

$$Fe^{3+} + 3\,HOH \rightarrow Fe(OH)_3 + 3\,H^+ \tag{2.5}$$

$$2\,S^{2-} + 3\,O_2 + 2\,H_2O \rightarrow 2\,SO_4^{2-} + 4\,H^+ + 16\,\text{electrons} \tag{2.6}$$

The net reaction products are Fe^{3+}, H^+, SO_4^{2-}, and energy in the form of electrons. The biochemistry of these net reactions will be discussed in greater detail in a subsequent section. The acid concentration of water draining from piles and mines commonly reaches a pH of 2.3 and occasionally reaches 1.8. This, in association with the high ionic strength of the water, has an extremely deleterious effect on the biology of receiving water. It is also very corrosive to metals and concrete.

2.3.3. Toxic Industrial Waste Minerals

Mineral pollution also originates directly from discharges of some manufacturing processes. The minerals can be either toxic or nutritious to living systems, but in either case they exert an influence on biological systems. Toxic ions such as cyanide, cadmium, mercury, chromium, and arsenic are often discharged from factories into lakes and streams. Perhaps the water pollution resulting from air pollution can be considered in the same category since much of this is mineral and directly related to technology. In the case of suspended particulate air pollutants (i.e., nongaseous air contaminants), it is axiomatic that they settle out across the land and water surface. A certain percentage of these particulates settles directly onto water surfaces. Of the amount which settles on land, a fraction is flushed into receiving water by urban runoff. The case of water pollution by urban runoff is significant and complex and will be considered in greater detail.

2.3.4. Radioactive Minerals

The problem of water contamination by radioactive materials is viewed as a mineral pollution problem for several reasons. Although the specific concern is with the hazardous effects of radiation, the mechanisms by which the radioactive substances are incorporated biologically are dictated by the chemistry of the radioactive molecule (e.g., substitution of Sr^{69} for Ca in milk and bones). It is readily apparent that some radioactive contamination overlaps with the topic of water contamination by air pollutants since strontium pollution originated primarily from air contaminated by nuclear explosions. However, a considerable amount of radioactive water contamination is reported to be a direct result of discharges by nuclear processing and power plants including ships. It also is worth mentioning that radioactive organic molecules (with the exception of recalcitrants) can be converted to minerals in the ecosystem. The technology required to remove radioisotopes from the environment will depend on the chemistry of the molecules, not on the fact that they are radioactive.

2.3.5. Mercury Pollution

Mercury contamination of the environment has received considerable attention recently because of its inherent toxicity to living forms. Mercurial salts and organic mercurial derivatives have long been used as bactericides, algicides, and fungicides because of their toxic properties. Many will recall having used mercurochrome (merbromin) as a topical antiseptic to prevent infection in an open cut or wound. Organic mercurial bactericides and fungicides have found wide usage in paint formulations and as slimicides in paper or other manufacturing processes where bacterial slime production is troublesome. Large quantities of mercury are also used in the chlorine–alkali industry as the cathode in the electrolysis process to produce sodium–mercury amalgam. Some of the discharges from this process contain mercuric chloride and metallic mercury (Wilber, 1969).

The toxicity of mercury for humans was dramatically demonstrated when, in 1953, more than 100 Japanese in the fishing villages surrounding Minamata Bay were struck with a mysterious neurological illness. Over 50 people died from "Minamata disease," which was later shown to be caused by lethal accumulations of methyl mercury in fish and shellfish eaten by the villagers. A search ultimately traced the source to a plastics factory that dumped used methyl mercury catalyst and other organic mercurials into Minamata Bay. The ability of certain aquatic organisms to concentrate and accumulate mercury to lethal proportions increases the hazard of dumping mercurials into water. Toxicity is also known to be potentiated by the presence of copper ion in water.

In addition to the concentration phenomenon, many aquatic microbes as well as higher organisms are extremely sensitive to low concentrations of mercurials. Some of the reported lethal concentrated are 5 parts per billion (ppb) for *Daphnia*, 30 ppb for the phytoplankton alga *Scenedesmus*, 200 ppb for the bacterium *Escherichia coli*, and 20 ppb for salmon. Lethal dosage varies with the mercury derivative and with environmental variables.

Mercury in living tissues appears to be primarily methyl mercury, CH_3Hg^+, or dimethyl mercury, $(CH_3)_2Hg$. Methylation of mercury in the environment is due to bacterial activity—particularly that of anaerobic bacteria at the bottom–water interface—and this is the form in which it is accumulated by organisms (Wood *et al.*, 1968).

Naturally occurring ores of mercury are dispersed in rocks, soil, air, and water. Cinnabar (mercuric sulfide) is the most common mineral form. Sedimentary rocks, organic-rich shale, sulfur-containing coal, and crude petroleum often have a high mercury content ranging up to 20,000 ppb. Some tar fractions of petroleum contain up to 500,000 ppb mercury. All this points out that some of the mercury pollution could arise from natural causes such as erosion or as secondary contamination along with pollution related to handling and combustion of coal, fossil fuel, etc. The potential hazard as an air pollutant appears to be greater than as a water pollutant. The difficulty in assessing historical sources of mercury pollution is related to lack of background analytical data from which comparisons can be made (U.S. Geological Survey, 1970). It is, however, interesting to note that high mercury concentrations under natural conditions are nearly always associated with minerals of biological origin (e.g., coal-, petroleum-, and sulfur-containing sediments), a historical factor which is not unrelated to the current pollutional involvement. This is due to the propensity mercury has for reacting with either free sulfide ion or an organic compound of sulfur.

2.4. POLLUTION BY RECALCITRANT MOLECULES

Another general area of current concern is with the biological effects of recalcitrant molecules. These are synthetic compounds which are not generally amenable to degradative enzyme attack and therefore remain indefinitely in the environment. Some of these compounds have marked toxicity to biological systems in extremely low concentrations. Examples of this type of pollutant are chlorinated hydrocarbon pesticides and detergents of the aryl and alkyl sulfonate type. Some of the recalcitrant molecules such as polyethylene and fluorinated hydrocarbons are nontoxic but still have pollution implications because they will not decompose. They appear to be destined to clutter the environment indefinitely and therefore constitute an environmental chemical "sink" as they remove carbon and other chemicals from the natural pool of cycling chemicals for an indefinitely long period. The recalcitrant pollutants will be discussed in greater detail in Chapter 9.

2.5. HEAT POLLUTION

Finally, I would add as a major consideration, heat contamination. This is also a by-product of all biochemical reactions but is of critical importance because of current technology. Of course, everyone is aware that increased temperatures accelerate rates of chemical and biochemical reactions. Biological systems have optimal temperatures at which individual enzymes function, and deviations from the optimum will markedly alter the reaction. In general, a marked lowering of temperature will decrease the reaction rate, whereas a marked increase in temperature will increase the reaction rate until the enzymes are physically denatured. If the reaction which the enzyme catalyzed is essential to the organism, it will result in death of the organism. Enzymes generally do not tolerate temperature increase more than a few degrees above their optimum.

Chemical reaction rates generally double with every 10 C rise in temperature (Q_{10}). Biological systems are highly organized and integrated chemical systems that contain some components which are easily denatured by heat. Although rates of biological reactions usually increase with increased temperature, the integrated sequences of reactions do not follow the Q_{10} rule for simple chemical reactions. Biological systems (cells) have an optimum temperature range within which the overall system functions "best." In many cases, the death rate increases sharply as temperature surpasses the optimum by a few degrees. Biological activity ranges from -10 C to about 100 C, with relatively few species active at the extremes. Each species has its own optimum range, which is rather narrow for non-warm-blooded organisms. On a biochemical level, the effect of temperature can be interpreted in terms of energy required to break hydrogen bonds (see Chapter 6).

Cooling water from nuclear reactors is reported to be between 10 F and 30 F higher at the outlet depending upon plant design and operation. Fresh water withdrawal requirements for the power industry were about 62 billion gallons per day (bgd) in 1965 and are estimated to be about 134 bgd in 1980, with much of the increase shifting to nuclear reactor use in contrast to fossil-fuel power generation (see Chapter 3). This implies that approximately 134 bgd of cooling water will be heated an average of 20 F and put back into the freshwater lakes and streams by 1980. For comparative purposes, this is roughly equivalent to 1 percent of the volume of Lake Erie per day. However, rapid and complete mixing will not occur, and we can anticipate localized areas with extreme temperature increases accompanied by drastic ecological changes.

From an ecological viewpoint then, heat will result in killing certain indigenous organisms while accelerating the growth of more heat-tolerant ones, and a general shift in populations will occur.

The significance in raising the temperature of receiving waters even a few degrees lies first in the predictable effect of shifting the optimum temperature for certain species while making it more favorable for other species. One cannot, however, predict at this time, except in gross generalities, which species will proliferate and more importantly what their biochemical activities will be under the new condition. Secondly, some species will die out altogether. If enough heat is added, sufficient evaporation could occur to endanger shallow ecosystems and also to influence weather conditions.

Chapter 3

WATER IN PERSPECTIVE TO POPULATION AND POLLUTION

Various people have different impressions as to what water is and how it behaves both *physically* and *chemically*. These differing impressions are largely the result of the individual's interest in water, his background and training, and so on.

For example, a civil engineer may view water as billion gallons behind a proposed dam. The water will exert a calculated pressure at the base of the dam, and therefore the dam must be constructed in a prescribed manner to give the necessary strength.

A biologist or chemist may always visualize water as H_2O. H_2O splits and adds to a double bond to form a glycol or hydrolyzes a dipeptide bond to form two amino acids.

A layman may think of water merely as something to drink, bathe in, or wash his automobile with.

A politician may visualize water in terms of a tax levy, bond issue, or a potential political problem.

The factor which causes most of our difficulty in dealing with water is failure to recognize the other person's point of view, or to consider it in proper perspective, and therefore we are not always sympathetic toward individual problems. Secondly, if all "experts" had a broad perspective, many of the ridiculous and costly mistakes of the past would not have been made.

We shall now consider water in its broad perspective and then turn our attention to areas more related to biochemical ecology.

3.1. HYDROLOGIC CYCLE, WATER AVAILABILITY, AND USE

The total amount of water on earth has been estimated as 10^{15} acre-feet (Wollman, 1962). Of this, 97 percent is ocean and therefore nonpotable, and only 3 percent or 33×10^{12} acre-feet is fresh water.

Of the fresh water, 89 percent is not readily available, as illustrated in the following breakdown:

> 75 percent of fresh water is ice in glaciers or polar ice
> 14 percent is ground water at depths between 2500 and 12,000 ft
> 11 percent is ground water at depths less than 2500 ft
> 0.03 percent is in lakes
> 0.03 percent is in rivers
> 0.06 percent is present as soil mosture
> 0.035 percent is present in the atmosphere at any given time (Because of recycling, 7.7 times this amount is annually over land and 30 times this amount is annually over water.)

The annual distribution of moisture available through the atmosphere is shown in Figure 1. The available freshwater resources appear small on a percentage scale, but they are probably adequate to support the population until at least the year 2000. One major problem in this regard is the uneven distribution of population, and localized water shortages are anticipated. Localized shortages result from a drain by major urbanized areas which also

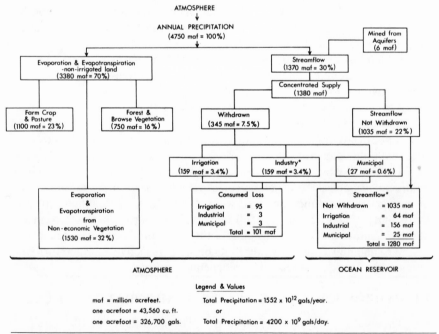

Figure 1. Over-all distribution of precipitation in the United States. (From Wollman, 1962.)

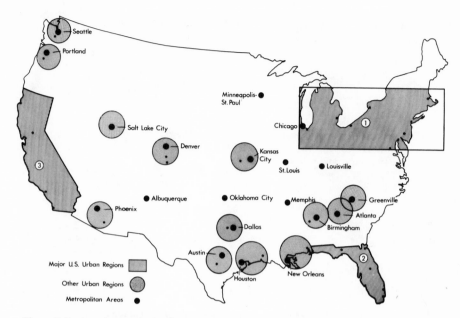

Figure 2. Projected urbanization in the year 2000. (From Pickard, 1967, and *Chicago Tribune*, March 26, 1967, Section 3A, p. 1.)

require large industrial supplies for manufacturing. This problem is compounded because the very areas requiring the most water are also the greatest sources of pollution. Population projections for the year 2000 indicate that 77 percent of 312 million people in the conterminous United States will occupy about 11 percent of the land area and that 95 percent of the population will live in urban environments (Pickard 1967). Figure 2 illustrates the projected urbanization in the year 2000. For example, the entire Atlantic Coast region of the United States had about the same average precipitation (40 to 60 inches/ year), average runoff (5 to 20 inches/year), average evaporation rate (40 to 60 inches/year from Massachusetts to Florida), and average water balance surplus (up to 20 inches/year). When this area is divided into northern and southern regions having Norfolk, Virginia, as the midpoint, the northern region used 37,467 mgd whereas the southern region used 20,560 mgd in 1965 (U.S. Water Resources Council, 1968). The reason of course is primarily due to a higher population density in the northern region than in the southern region.

The total United States water use is presented in Table I for the year 1965 and is compared to the projected use requirements for 1980. The requirement has been subdivided into use type on a withdrawal basis and on a consumptive basis. Agricultural irrigation was the largest single use of water,

Table I. Estimated United States Water Use (mgd) and Projected Requirements*

	Withdrawal		Consumptive use	
Type of use	Used 1965	Projected 1980	Used 1965	Projected 1980
Rural domestic	2,351	2,474	1,636	1,792
Municipal (public-supplied)	23,745	33,596	5,244	10,581
Industrial (self-supplied)	46,405	75,026	3,764	6,126
Steam-electric power				
Fresh	62,738	133,963	659	1,685
Saline	21,800	59,340	157	498
Agriculture				
Irrigation	110,852	135,852	64,696	81,559
Livestock	1,726	2,375	1,626	2,177
Total	269,617	442,626	77,782	104,418

*From U.S. Water Resources Council, 1968.

Table II. 1965 Industrial Water Use (bgd) in the United States*

Type of industry	Withdrawal	Processing	Cooling	Other
Manufacturing	40 (90)[a]	10.6	26.5	2.9
Mining and mineral processing	3.2 (11)[a]	2.07	0.9	0.3
Ordnance				
Construction	0.04			
Government military installations	0.6			

*From U.S. Water Resources Council, 1968.
[a]Values in brackets indicate gross use and account for water recycling.

Table III. Tons of Water Required per Ton of Consumer Product (maximum)

Product	Water required
Living tissue per year (e.g., meat)	10
Brick	1–2
Paper	250
Fertilizer	600
Corn or sugar	1,000
Wheat	1,500
Rice	4,000
Cotton fiber	10,000

and over half of the water was lost or consumed. Sixty-two billion gallons per day were used for power generation, but little of this was consumed. This did, however, represent heat pollution. Industrial uses totaled 46 bgd, which can be subdivided as shown in Table II. Eighty-eight percent of manufacturing water was utilized by the following industries: food and related products, pulp and paper, chemical, petroleum and coal products, and primary metals.

Values presented in Table III illustrate the estimated amount of water required to produce certain consumer goods (Maxwell, 1965).

United States water use requirements have been projected with regard to withdrawal and consumptive use (U.S. Water Resources Council, 1968). Withdrawal uses are primarily for domestic, industrial, steam-electric power, and agricultural purposes, as indicated in Table I. Figure 3 illustrates that per capita use of water by individual systems is expected to increase significantly, whereas the per capita use of water through public distribution systems is expected to increase quite modestly. This is primarily due to an adequate water supply to those now being served by municipal systems and the generally inadequate supply to those who are not. Figure 3 can be misleading, and it should be stressed that total withdrawal will increase significantly because of increase in population. Also, nonconsumptive use will increase the pollutional load of water.

One of the largest users of water is, and will continue to be, the power industry (See Table I). This is primarily a nonconsumptive use but could contribute significantly to the heat pollution problem. Figure 4 shows the projected electric power generation, over 90 percent of which will be supplied

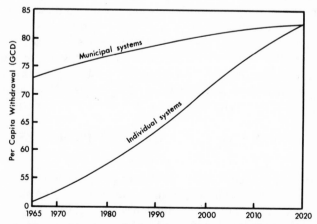

Figure 3. Actual and projected domestic water withdrawal per capita for the years 1965–2020. (From U.S. Water Resources Council, 1968.)

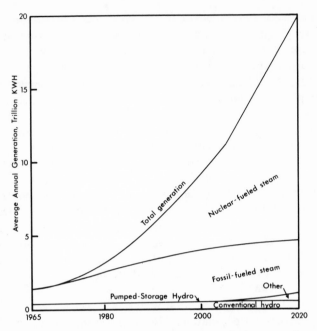

Figure 4. Projected electric generation by types of prime mover for the years 1965–2020. (From U.S. Water Resources Council, 1968.)

by fossil-fueled and nuclear-fueled plants. The amount of water required for cooling purposes is expected to follow the pattern for power generation.

Figure 5 illustrates United States regional indices of projected fresh-water withdrawals and consumptive uses. In compiling these data, the U.S. Water Resources Council (1968) cautions that (1) use figures include use of mined groundwater which is prevalent in the Southwest, (2) use requirements include some saline withdrawals by self-supplied industries, (3) use figures do not include phreatophyte or other channel losses which are particularly important in arid regions, (4) the water use data do not include requirements for instream uses, and (5) the data do not reflect seasonal variations.

3.2. POPULATION GROWTH AND ITS BY-PRODUCTS

Pollution, in overview, is both produced by people and affects people, which makes it a social consideration. The production of pollution by popula-tions can be considered separately from its effects. Production has two facets. One is the inevitable by-product production which is governed by natural law. It therefore cannot be stopped or cured without reducing the population. By-product accumulation can be altered somewhat, although it is debatable whether treatment measures would be significant without decreasing population growth. This type of pollution will be referred to as first order pollution. The second facet will be referred to as second order waste accumulation and implies waste that is formed as a result of man's technological activities. The amount of secondary waste is also directly related to population size. More importantly, it is related to the type of activity and energy expenditure of the population. This facet of pollution production is therefore amenable to alteration by changing the type of energy expenditure. In this case, we have a choice of either slowing down the secondary production or adding more treatment technology to replace missing segments in the elemental cycles that govern the balance of nature.

Basically, then, first order pollution is of a type which is governed by natural or physical laws and is nonnegotiable within the context of a living population. Second order pollution is of a type which is governed by agreements and decisions within the context of a living population. This is pollution based upon technically artificial laws. It is negotiable and can be prevented.

A large proportion of environmental pollution can be attributed to direct effects of the technological revolution, but in the ultimate analysis it is the fault of people, not the fault of technology. That is to say, we also have the

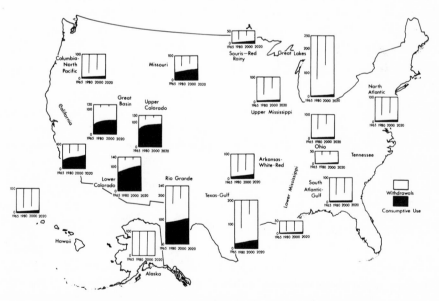

Figure 5. Regional indices of projected water withdrawals and consumptive uses for the years 1965–2020. Estimated average supply equals 100. (From U.S. Water Resources Council, 1968.)

technological capability to remove those pollutants produced by technological exploitation, but we have not yet chosen to make use of our technological
expertise. This gets into the realm of socioeconomic considerations and is
somewhat divorced from biochemical ecology. Nevertheless, the public
could demand that segments of the population stop their contaminating ways
altogether or demand that they add the technical capability of pollution
abatement directly to the process which produces it.

3.2.1. First-Order Pollution

Since first-order pollution is part of the life process and the pollutional
effects are related to accumulation of biological waste products (excreta) in
the environment, it becomes a function of concentration and therefore of
total numbers of people. Figure 6 is a chart showing United States and world
population projections from the present 3.36 billion through the year 2040,
assuming the present growth rate of 2 percent. An anticipated drop in the
world birth rate from the present 33.2 to 25.1 per 1000 people, accompanied
by a drop in the death rate from 12.8 to 8.1 per 1000 people, would lower the
growth rate to 1.7 percent by the year 2000.

The amount of human biological waste varies widely with individuals
but averages approximately 1.25 g of solid, on a dry weight basis, per kilogram
of body weight per day (Albritton, 1952). Assuming a United States population of 200 million and an average body weight of 68 k (150 lb), we are
presently excreting 40 million pounds of solid waste per day in the United
States. This excreta consists of about 25 percent nitrogen as NH_3, and a
large proportion of it finds its way into surface water through a sewerage

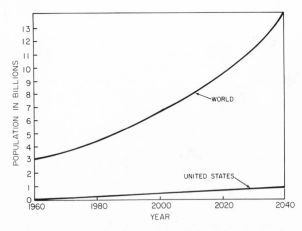

Figure 6. Actual and projected United States and world population for the years 1960–2040.

system or other means. Every equivalent of nitrogen will require 4.5 equivalents of oxygen to oxidize it completely to nitrate (NO_3^-). This is one of the objectives of aerobic waste treatment. On a weight-to-weight basis, 10 million pounds of nitrogen will demand 51 million pounds of oxygen per day to carry out the complete oxidation, and it will result in enriching the environment with 36 million pounds of nitrate per day.

The solubility of oxygen in water at 20 C is about 8 g per 1000 liters, and consumption of 51 million pounds of oxygen (23×10^9 g) is equivalent to saturating the assimilative capacity of about 7×10^{13} liters of receiving water or nearly 6.6×10^7 acre-feet of water per day. The total world supply of available fresh water is estimated to be on the order of 4×10^{12} acre-feet, and only 25 percent of the excreta from the United States (i.e., the nitrogenous portion) was used in the calculation. Extrapolation to present world population using a factor of $20 \times$ indicates that all of the oxygen dissolved in 1.3×10^9 acre-feet of fresh water would be required to assimilate the world's human nitrogenous excreta each day. This is within a factor of 3000 of using the total fresh water available, or a factor of 1000 if we assume that 50 percent of the remaining nonnitrogenous fraction of excreta, which contains organic carbon, phosphate, sulfur, and other minerals, requires an amount of oxygen equivalent to that for NH_3 consumption.

The above calculations, although argumentative, serve to illustrate some of the relationships between population and the demand for dissolved oxygen to assimilate its excrement.

The calculations are not accurate for several reasons. The amount of excreta for each individual depends upon food intake, and United States values are not demonstrative of those in undernourished countries. We have no way of estimating the amount which actually finds its way into fresh water as compared to the ocean, but it is reasonable to assume that relatively little gets into ice. Surface water is continuously reaerated by diffusion and mixing and also through photosynthetic production of oxygen by algae. On the minus side of the balance, we would have to consider oxygen-consuming waste products other than human excreta that enter sewers and surface water.

Increased use of kitchen disposal units has increased the domestic sewage load tremendously in the past 10 years, and the volume of domestic sewage is increasing in proportion to population. The projected municipal waste disposal requirements are shown in Figure 7 in terms of biochemical oxygen demand (BOD) of the oxidizable organics and also in terms of population numbers.

The average BOD value in the United States is about 0.15 lb BOD per capita per day. This amount of sewage oxygen demand is present, on an average, in 160 gal of water per capita per day.

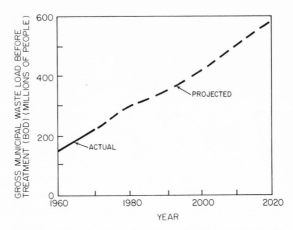

Figure 7. Actual and projected municipal waste disposal requirements for the years 1960–2020. (From U.S. Water Resources Council, 1968.)

Growth of most populations is theoretically a geometric progression. That is, 2 individuals produce 4, 4 produce 8, 8 yield 16, etc. Human population growth patterns deviate somewhat from this pattern for a variety of reasons such as changes in food supply, disease, multiple births, and the "pill." However, populations of single-celled organisms such as bacteria do follow a theoretical geometric progression up to a point, provided the environment is ideal for reproduction. If increasing numbers of bacteria are plotted against time, we observe an exponential increase.* This is illustrated in Figure 8, which shows a nearly asymptotic rise when numbers double every half unit (e.g., half hour). Figure 8(B) illustrates that when the same values are plotted as logarithms to the base 10 vs. time, a linear response is observed. When large numbers that would not fit into the plot shown in Figure 8(A) are plotted on a condensed scale as shown in Figure 8(C), a sigmoidal or S-shaped curve is developed. The point in considering an S-shaped population growth curve is that the population does not continue to increase indefinitely. It levels off when the death rate equals the generation rate of individuals. The population of course declines when the death rate exceeds the generation rate and is generally called the decline phase of the growth curve. The two predominant factors which contribute to a decline phase in a closed system are depletion of nutrients and the accumulation of toxic by-products. In a large sense, the earth is a closed system in which nutrients are being depleted and toxic by-products are accumulating.

*Growth of individual birds and mammals, as contrasted to populations, has also been shown to proceed exponentially, but the specific growth rate decays exponentially as the animal ages (Laird, 1969).

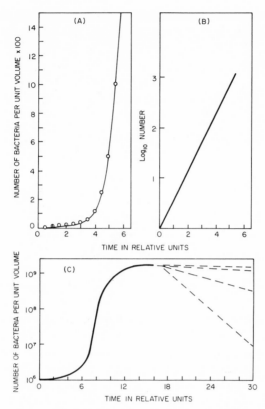

Figure 8. Growth curves showing increase in number of unicellular organisms vs. time. (A) Arithmetic plot. (B) Logarithmic plot. (C) Extended log plot.

3.2.2. Second-Order Pollution

This type of pollution can be referred to as our Gross National By-product. In terms of energy, waste materials accumulate in proportion to population numbers and their activity. The United States population is the most active in the world and therefore expends more energy per capita while accumulating more waste product per capita than other populations.

A comprehensive economic measure of this activity (energy turnover) is the gross national product (GNP). This is a measure of the total market value of all goods and services produced. GNP is compared to the population for the period 1910 to 2020 in Figure 9. What has not been generally considered by economists who use the GNP curves as an indicator of national

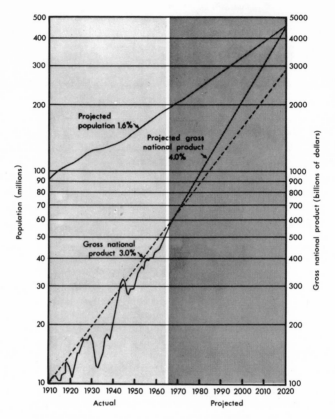

Figure 9. Historical and projected population and gross national product (1954 dollars) for the years 1910–2020. (From U.S. Water Resources Council, 1968.)

well-being is the inevitable by-product accumulation from energy expenditure. In this sense, economic "health" as measured by present indicators is not indicative of national "health."

An economist, A. A. Berle (1968), recognized that junk piles and other wastes were not accounted for in calculations of GNP, nor were what he calls "disproducts" added to calculations of national capital consumption. An example of a disproduct was given as the cost of future cancer generated by this year's output of tobacco—a cash value crop accounted for in the GNP. The author also points out that economists are basically accountants who do not make value judgments of this type, although he stresses the need for an index of social as well as dollar benefits.

Aside from biological excreta, the population in the United States at the present time is accumulating solid wastes at the rate of 4.5 lb per day per capita or about 1500 lb per man per year. Solid wastes are generally those wastes which do not go through sewage systems and consist largely of disposables of all kinds such as packages, bottles, cans, rags, and durable goods which were not so durable. Sixty-two billion cans were used in the United States in 1968 with a per capita consumption of 308 cans. Of these, over half are used as packages for one-way items and are not refillable and therefore would not be amenable to substitution by returnable containers.

In a 29-state study, litter was found to consist of paper packages and containers (60 percent), metal cans (13 percent), newsprint and discarded metal scrap, etc. (27 percent) (MacIntosh, 1971).

Much of the accumulation of solid waste is clutter and not pollution in the context we have been using unless one chooses to define it as aesthetic pollution. Accumulations of glass bottles are repulsive, for example, but will not directly influence living systems in an aquatic environment. However, the portion of solid waste that is biologically degradable does contribute directly to our water pollution load. No accurate figures are available as to what percentage of the total solid waste this represents.

Several years ago, solid waste accumulations did not matter to most people because they lived at some distance from dumps, riverbanks, railroad tracks, and factories. With increased land values, this is no longer true. People now live among the refuse, and it is becoming increasingly more difficult to physically remove it. Although the service demand is up, this type of job demand is down. No one wants to collect trash! It seems evident that our political-economic-social incentive system of enterprise will either get a compete overhaul or we are destined to wallow in our waste.

PART II

BIOCHEMICAL CONSIDERATIONS

Chapter 4

BIOCHEMICAL ASPECTS OF WATER POLLUTION

Water pollution when considered in its broadest context is a by-product of human population, and its significance is in what effect it has directly or indirectly on living populations. The pollutants are chemicals (except heat) that interact with living cells, which is the subject of biochemistry. Biochemical reactions alter the compounds chemically and therefore physically because physical properties depend upon chemical configuration. The net result is an alteration in environmental quality.

Biochemistry of water pollution is no different from other biochemical considerations, and one could say that basic biochemistry is adequately covered in any good text on the subject. The most active biochemical agents in the aquatic habitat are the microorganisms—primarily the bacteria and algae—and their activities are described in most microbiology texts. However, the metabolic activities of organisms acting on pollutants are generally not considered, so a discussion of biochemical reactions of pollutant molecules is warranted. The reactions of greatest relevance to water pollution are those concerning biodegradation and chemical conversions that either stimulate or inhibit other cells. Some pollutants can also act directly as toxicants or nutrients without undergoing a degradation or chemical transformation, and these should also be considered. On the other hand, some become toxicants or nutrients only after being acted upon biochemically. There is relatively little reason to consider the metabolic pathways of biosynthesis except as such reactions are pertinent to pollutional considerations.

There are several ways of approaching the myriad of biochemical reactions and pathways in addition to environmental controls that influence the rate and extent of reactions.

Theoretically, pollutants consist of all known chemicals synthesized naturally or synthetically. One must therefore consider general classes of compounds such as polysaccharides, sugars, proteins, amino acids, organic acids, polyols and alcohols, aliphatic and aromatic hydrocarbons and their derivatives, lipids, nucleic acids, and minerals.

The approach followed in this monograph will be to describe the bio-chemical reactions by subtopic and relate these to identifiable problem-area ecosystems. For example, all treatment of organic wastes is presently based on the biological oxidations catalyzed by microorganisms. If the environment is aerobic, aerobic organisms will predominate because of their ability to carry out oxidative metabolic reactions.

Biochemical oxidations are energy-yielding reactions, and it is the energy liberated during the oxidative reactions that is utilized by organisms for synthesis of new cells. Much of the pollutant chemical is then transformed into new cell mass. An oversimplification of the process is shown below, where a variety of terms used by various groups of people imply the same overall process, i.e., that pollutants are used as nutrients by microorganisms and oxygen is consumed in the process:

pollutants + microorganisms + $O_2 \rightarrow$

$$\text{oxidized pollutants + organisms + } H_2O \qquad (4.1)$$

nutrients + microorganisms + $O_2 \rightarrow$

$$\text{oxidized nutrients + organisms + } H_2O \qquad (4.2)$$

$$\text{sewage + flocs + } O_2 \rightarrow \text{treated sewage + more flocs + } H_2O \qquad (4.3)$$

waste + activated sludge + $O_2 \rightarrow$

$$\text{treated sewage + more sludge + } H_2O \qquad (4.4)$$

The terms in these word equations differ because the jargon of waste treatment plant operators and engineers has been different from that of chemists or biologists. The process of biochemical oxidation which has been described can proceed in any lake or stream just as readily as in a treatment plant. There are no facilities to aerate (add oxygen to) large bodies of water. Therefore, it will not be long before organic contaminants which may be dumped directly into lakes without prior treatment will demand most of the available dissolved oxygen, which has a solubility of about 8 ppm at 20 C. When this is allowed to happen, other living things in the lakes which also require dissolved oxygen will perish. This is often the cause of fish kills, and it illustrates the importance of oxidizing waste in a treatment plant where we are able to replenish the dissolved oxygen by pumping air into the system as it is required.

As yet we have not said anything about how the microorganisms actually carry out these biochemical oxidations, which are known collectively as metabolism. These processes depend upon the individual kinds of micro-organisms present and upon the individual kinds of waste substances being treated. For example, the oxidizable material in domestic sewage is usually different from that found in industrial wastes, such as from milk processing plant effluents.

There are ecosystems where oxygen is absent. In this situation, higher forms of life do not exist and protists are the only cells present. Anaerobic microcosms are very common in the aquatic environment, and they are essential to chemical transformation of pollutants. The biochemical reactions of anaerobic organisms that are of significance to pollution abatement fall into the following general categories: (a) anaerobic respirations, in which sulfate, nitrate, or carbon dioxide substitutes for oxygen; (b) fermentations, in which a carbohydrate is converted to an organic acid or alcohol; (c) hydrolysis, in which polymers are degraded to their monomeric units (hydrolytic reactions are also carried out by aerobes).

Many organisms have the facility to metabolize and grow in either the presence or absence of oxygen, and the reaction patterns of these facultative organisms will be characteristic of aerobic and anaerobic forms depending on whether oxygen is present or absent.

Certain basic biochemical considerations are common to all organisms. Nearly all biochemical reactions are catalyzed by enzymes that are highly specific for the chemical substrate being acted upon. There are nearly 2000 known enzymes, and no single cell or species of organism has all of them. Of the enzymes present in an individual organism, some are constitutive and others are induced in response to the presence of a substrate chemical. However, capacity for synthesis of either type of enzyme is a genetically conferred trait.

Several different species of organisms are usually required to produce the necessary enzyme complement needed to degrade the variety of complex chemicals present as environmental contaminants. Complex chemicals therefore will be sequentially degraded. As one enzyme acts on a pollutant it will be modified chemically, and then it is a substrate for a different enzyme. The potential exists for all types of microorganisms to be present in any aquatic or soil microcosm at any point in time. Various species of organism will proliferate or fail to proliferate depending upon whether they have the capacity to produce the necessary enzymes for the available substrates. In this manner, successions of organisms take place and certain species will predominate, although in the environment the considerations are much more complex. For example, several species can compete for a given substrate, and rates of reactions are altered by temperature, pH, light availability, oxidation–reduction potential, availability of nutrients, and other environmental variables.

Chapter 5

ECOLOGICAL CONCEPTS

An environmental niche can be considered as a subdivision of the biosphere characterized by a particular set of biotic and abiotic factors, and in a time sense no single group of organisms can be separated from the rest of the biota in an ecosystem. An ecosystem is a concept which cannot be measured or delimited directly. Nutrients and nutrition are primary factors in determining the kind of community which develops. Variety and concentration of nutritional substrates are primary factors in determining the community (Hungate, 1962).

All biochemical reactions under natural circumstances are related to the utilization of chemicals as nutrients by one organism or another for the purpose of cell growth (as energy or structural components) or to rid the cell of waste products.

Organisms have evolved in an aqueous menstruum. Acquisition or possession of the enzymes necessary to utilize available chemicals gives survival value to the organism. Organisms which have the capacity to adapt to a given set of environmental conditions will also have survival value. The scheme of evolving life patterns is such that organisms both act to modify their environment and are selected out by the continuously altered environment. The environment has become more suitable for more efficient organisms. One should keep in mind that it does not matter how the environment comes to be modified—the population of organisms will react and interact accordingly.

Speakman (1966) considers biology as "just a configuration of the natural elements" and argues on the basis of the chemistry of the elements that living structural macromolecules had to evolve from a limited number of elements restricted to a small section of the periodic table (i.e., H, C, N, O, Si, P, S). The primary basis for his point of view is a consideration of reactivity and stability of molecules within the life temperature range for liquid water. Biochemicals of necessity must be relatively stable at these temperatures, which provides for orderly sequential rate-limited reactions and a slow release of heat (entropy). This allows cells to metabolize without "burning up" or undergoing heat denaturation of hydrogen-bonded molecules (see Chapter 6). Further, the molecules must not be too stable, or otherwise reactions would either not proceed at all or would be too slow to accommodate life-sustaining energy transfer.

At any point in time, a finite amount of any chemical is available in the biosphere. Cycling of the chemicals is an absolute prerequisite for continued sustenance of life as we recognize it. Part of the overall pollution problem results from release of chemicals from an extrabiosphere reservoir, which then become available as nutrients which stimulate growth of organisms. It is often the subsequent biological activity which is recognized as an environmental problem. For example, exposure of pyrite from mines results in mine acid formation. Also, production of CO_2 from oil combustion ultimately results in stimulated plant (algal) growth.

Adaptation of organisms in the environment occurs in two general ways: (1) by modification, which is a temporary change in an organism's character (this may result from inducible enzyme formation), and (2) by genetic mutation, which is a permanent change in the hereditary mechanism of the organism and is transmitted from parent to daughter cell.

All cells seem to undergo spontaneous mutations for any given identifiable trait. The frequency of mutation under natural conditions appears to be in the range 1×10^{-4} to 1×10^{-10} depending upon the type of cell; i.e., in a population of growing (dividing) cells, a mutant will appear in every 10^4 to 10^{10} cell divisions. Mutation frequencies are given for a single recognizable trait (e.g., ability to utilize arginine as a nutrient). If all possible traits could be recognized, it might be possible to demonstrate that some type of mutation occurs each time a cell divides. It is also known that various environmental factors can induce an increase in mutation rate for a given trait from tenfold to 100,000-fold. These are known as mutagenic agents and include such agents as ultraviolet and X-rays, organic peroxides, nitrous acid, nitrogen mustard gas, and $MnCl_2$.

Many mutations on the cellular level are lethal, and the mutant does not persist; i.e., the mutation has negative survival value. Only mutants with positive survival value persist. In this context, organisms having high population density and short life cycles (rapid generation time) are favored with regard to natural selection and survival in a changing environmental niche.

Organisms have adapted to their environment, generally, by becoming increasingly complex with more specialization or parts.

The following remarks, taken from Brown (1958), although directed to animal and plant adaptation, seem to be applicable to all organisms and therefore relevant to the present discussion:

> Within single species, geographical variation seems to be correlated with two types of adaptations: *special adaptations*, in which the variation follows particular environmental variables, and *general adaptations*, in which the variation takes up the broad pattern (often centrifugal) independent of environmental differences within the species range. Both at and above species level, special adaptations appear to function to fit the organism or population to the particular features of the immediate environment. General adaptations are concerned more with internal organization and efficiency of the individuals or population, more or less independently of the details of the environment. Of course, the two types of adaptation differ as extremes of emphasis, and not as absolutes. Adaptations that permit a species or a group to extend its range permanently are in a different category, called *extension adaptations*, which may include special and/or general adaptations. Species-groups or genera normally contain one or more potent species in which general adaptation is prominent. The potent species represents a kind of "apical growing point" of general adaptation, and continues the line most likely to give rise to new major groups. The "lateral" component in evolution is represented by the more specialized species of the group. The differences between these are primarily due to *adaptive radiation* based on repeated *character displacement*.

Mutations and natural selection of populations of organisms are illustrated by comparison of the flora of activated sludge and anaerobic waste treatment processes, acid mine water, hot springs, solid waste landfills, etc., where shifts in the chemical composition of the pollutant load will result in a selective shift in the microbial population.

5.1. SYMBIOSIS

The term *symbiosis*, originally defined by DeBary (1897) to describe lichens, implies a constant intimate relationship between dissimilar species. The original usage of the term had no connotations of strict mutualism, which is bilateral advantageous symbiosis. The symbiotic association is usually dependent upon environmental factors. As the term has come to be used, it generally implies a certain degree of specificity and permanency, although the implication of a state of permanency must be made with caution. Variations on the symbiotic theme are (a) *commensalism*, where only one of the symbionts profits, or there is neither advantage nor disadvantage; (b) *parasymbiosis*, where there is no physical contact and no mutual harm. It is a mutual physiological relationship sometimes called *parabiosis*; *parasitism* is the overt exploitation of one symbiont by the other, leading to injury or death.

The value of these distinctions is to widen the concept of organism in the natural state to include heterogeneous systems (or populations) overriding the limitations of genetic uniformity and to supplement the concept of structural unity by that of functional unity or functional field (Gregory, 1951). Henry (1966) has gone on to say "therefore just as we consider the complex multicellular organism whose unaccounted numbers of different types of cells interact for the good of the whole and are adjudged to belong to a single entity, so we can draw analogy to and appreciate the oneness of symbiotic pairs or larger groups and envision them in true relationship to their environment."

Three types of resultant symbiotic effects can be distinguished in mixed cultures or populations: (1) inhibitory or stimulatory effects on growth rate, (2) quantitative effects on biochemical activity such as an increase or decrease in metabolic intermediate or waste products, and (3) qualitative biochemical effects leading to a result that neither symbiont could produce alone. This effect is called *synergism*.

Synergism as a concept is somewhat akin to *syntropism*, which is the term applied to mutual growth stimulation. Quite often, one organism can nutritionally utilize a substance from the environment which is normally inhibitory to another. This is called *passive stimulation*. An example of this type of symbiosis is the production of methane from ethanol. What was heretofore considered to be *Methanobacillus omelianski* has been shown to be two different organisms. One, called the S organism, is a motile rod that produces H_2 from ethanol as shown in reaction (5.1), and the H_2 is autotoxic to the cell:

$$CH_3CH_2OH \xrightarrow{H_2O} 2 H_2 + CH_3COOH \qquad (5.1)$$

The other bacterium is a nonmotile rod designated as *Methanobacterium* strain M.O.H. and uses the H_2 to reduce CO_2 to methane, as shown in reaction (5.2):

$$4 H_2 + CO_2 \rightarrow CH_4 + 2 H_2O \qquad (5.2)$$

thereby allowing the S organism to proliferate (Bryant *et al.*, 1967).

Symbionts often derive ecological advantage over nonsymbionts. This thought considers biological specificity. Specificity is that property of two interacting systems which permits one to interact with the other with some degree of selectivity. Each organism forms a more or less critical portion of the environment of the other. Symbiosis influences evolution of organisms and in one sense tends to encourage retrogressive evolution by more specialized adjustment leading to decreasing adaptability leading to an increase in degree of specificity. Increasing mutual dependence leads to increasingly complex forms of intergration between symbionts and therefore decreasing flexibility and increased fastidiousness and specificity. An increasing body of data has led to the suggestion that eukaryotic cells have evolved from a symbiotic association of prokaryotic cells (Sagan, 1967; Cohen, 1970). This would lead to greater complexity and may indicate that man as well as all higher organisms are social entities resulting from a combination of the shared genetic equipment and metabolic systems of several evolutionary branches. Although the components may have undergone retrogressive evolution, the resultant form would be more highly evolved. Specificity is therefore concerned with a pattern of adjustment of the symbiotic pair.

The greater the number of factors which interact and the more selective one of them becomes, the smaller the probability of achieving the required combination for complementariness—hence the higher the degree of symbiotic specificity. Specificity is an expression of the mechanisms which determine why and how certain organisms are frequently found in association with others and why they are restricted in their use of a living environment to these organisms. As more and more of a symbiont's needs are satisfied by its partner, selective pressure on it decreases. Structure and functions no longer required degenerate, since natural selection does not operate against genetic changes which damage useless somatic attributes (Dubos and Kessler, 1963).

The common denominator in all of our deliberations is water. Perhaps it would be sensible to begin with a consideration of what water is and how it is involved with the subject at hand.

Chapter 6

WATER, ITS PROPERTIES, BIOCHEMISTRY, AND BIOLOGICAL IMPLICATIONS

6.1. WATER CHEMISTRY

In order to examine the role of water in living processes and as a solvent, it is necessary to consider its chemistry. Water has many unique properties, and some understanding of them is essential to an evaluation of its functions and versatility as a chemical in a cell and in the environment.

Although water is generally thought of as H_2O, which is composed of two atoms of hydrogen bonded chemically to one atom of oxygen, it is in nature composed of a mixture of deuterium oxide (D_2O, H_2^2O), tritium oxide (T_2O, H_2^3O), and H_2O—deuterium and tritium being isotopes of hydrogen. In addition, three isotopes of oxygen are present: O^{16}, O^{17}, and O^{18}. These six isotopes can be combined in 18 different combinations. There are also 15 different ionic forms in which these 18 combinations can exist, giving a total of 33 different substances which make up "average pure water," ordinarily represented as H_2O. Of course, the heavy isotopes are present in relatively low concentrations; e.g., the deuterium content of water is approximately 200 parts per million parts (ppm) of water, the O^{18} content is about 1000 ppm of water, and smaller quantities of tritium and O^{17} are present (Buswell and Rodebush, 1956; Choppin, 1965). However, these concentrations may be quite high with respect to biological systems, since some chemicals are known to exert a marked effect on metabolism of certain organisms in concentrations well below 1 ppm (e.g., vitamin B_{12}, chlorinated pesticides).

The two hydrogen atoms of water are bonded to oxygen at an angle of about 105°, but this angle may bend or vary considerably in the water phase (Pauling, 1960, p. 110). Hydrogen has a single electron and therefore no enclosing shell. It is able to attach itself to other atoms by means of its electron (valence bond) and also to associate strongly with oxygen atoms in

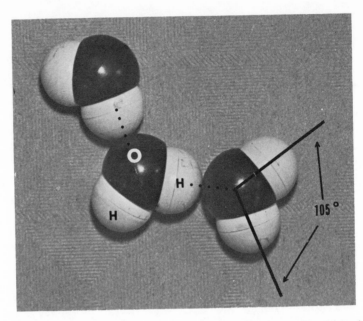

Figure 10. Photograph of three models of H_2O molecules showing arrangement of two H atoms (white) in association with one O atom (black). The H from one molecule orients with the O from a separate molecule. The intermolecule association between H and O is termed a hydrogen bond.

other water molecules by virtue of the attraction of its unoccupied positively charged "side" for the negative side of a second molecule, as illustrated in Figure 10. This attachment is known as hydrogen bonding. The hydrogen (H) bond is a relatively weak bond, having about 5 percent of the strength of a covalent carbon-to-carbon bond. Its properties are generally considered to be intermediate between covalent and ionic bonds. The bond energy has been calculated to be approximately 10 kcal per mole of water or about 5 kcal per mole of hydrogen bond, as compared to 85 kcal per mole of C—C or 100 kcal per mole of C—H. This implies that a relatively small activation energy is involved in the formation or rupture of a hydrogen bond. It should be kept in mind that although hydrogen bonds are weak, sufficient numbers give a great amount of collective bond strength to compounds. A simple illustration is the adherence between two pieces of wet paper or two wet glass slides compared to that between two pieces of dry paper or two dry glass slides.

Figure 11 depicts two hydrogen atoms, each from an individual water molecule, hydrogen-bonding to the oxygen of a third water molecule. The hydrogen-bond influence results in bending to an angle of 109° in the two

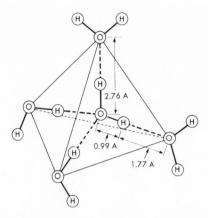

Figure 11. Drawing showing the geometry and arrangement formed by H_2O molecules hydrogen-bonded to each other. (From Snell *et al.*, 1965.)

covalent hydrogen atoms. The net effects of hydrogen-bonded water can be envisioned as four hydrogen atoms bonded to a single oxygen with a spatial distribution such that the hydrogen atoms form the points of a regular tetrahedron at angles of 109°, with the atom in the center. Additional HOH molecules could also be added and hydrogen-bonded in this configuration to form either a random-sized polymer or a crystalline lattice array. If the oxygen sphere at the center of the tetrahedron is represented as a node surrounded by four hydrogen bonds, as depicted in Figures 12 and 13, it can be shown that a solid crystal, consisting of 20 nodes (Oxygen atoms), can be built up

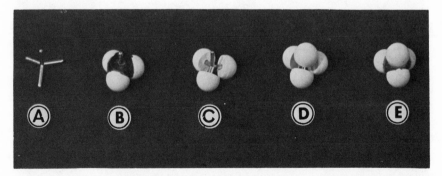

Figure 12. Models illustrating the tetrahedral arrangement of water molecules. (A) Four hydrogen bonds converging at a node representing oxygen. (B) Three atoms (white) around one O atom (black); a position is available for a fourth H atom. (C) Three H atoms around the O node; the O atom is removed from the model for simplicity. (D) Four H atoms around an O node. (E) Four H atoms around an O atom.

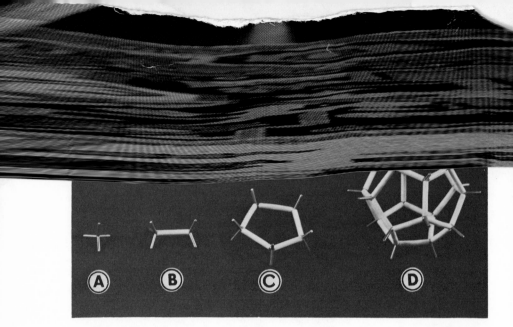

Figure 13. Models illustrating the formation of crystalline water having a pentagonal dodeca-hedral structure. (A) A tetrahedron of four H bonds around an O node. (B) Two tetrahedra associated by an H bond. (C) Five tetrahedra associated by five H bonds forming pentagonally arranged pseudocrystals of water (polymer). (D) A "water cage" consisting of H_2O arranged as a pentagonal dodecahedron.

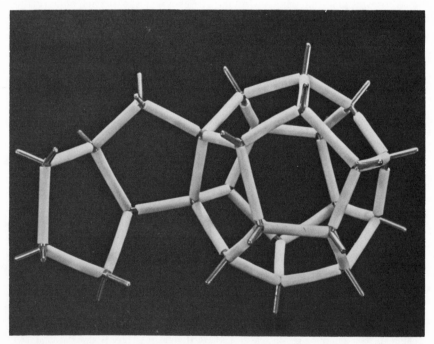

Figure 14. Model illustrating that a pseudocrystalline array of water molecules can "build up" to form larger aggregates than the dodecahedron shown in Figure 13.

will take the shape of a regular pentagonal dodecahedron, sometimes
ed to as a "water cage."

ditional hydrogen-bond bridging with more water molecules will
in the build-up of large crystalline lattices of the type visualized in
e 14. This is considered to be the arrangement of H_2O in ice, although
investigators disagree as to the exact bond angle (Runnels, 1966;
Eyring and Jhon, 1966). Since bond angles dictate the basic configuration,
it could be either six-sided or five-sided (Klotz, 1962). This crystal can be
visualized as a bridge in which a certain amount of inward pressure exists
If we add pressure, the lattice will collapse. Pressure can be added in the
form of heat or thermal agitation of molecules, compounds which have a
greater attraction for either the oxygen atom or hydrogen atom forming the
hydrogen bond (e.g., salts, urea) or actual mechanical pressure (e.g., ice
skating, wire cutting a block of ice). This is manifested as melting.

Hydrogen bonding can explain many of the anomalous properties of
water. Normally, one expects an increase in boiling point and a decrease
in melting point as molecular weight increases in a chemically related series
of molecules. (It should be kept in mind that boiling and melting points are
measures of specific heat capacity and heat of vaporization of substances
and therefore are indicators of intermolecular attractive or repulsive forces.)
Table IV compares water to low molecular weight alcohols in this respect
(water being considered as an alcohol in which hydrogen replaces carbon).
On a comparative basis, water would be expected to boil at approximately
55 C and not at 100 C and would be expected to freeze at about −45 C and
not at 0 C. One major environmental significance of hydrogen bonding of
water is that it reaches its maximum density at 4 C above its freezing point.
This means that frozen water is less dense than slightly warmer water, and it
will therefore freeze at the surface of a lake rather than at the bottom. If this
were not the case, all aquatic life would be different than that known. A
similar comparison is made in Table V when the oxygen of water is replaced
by other atoms of period VIA from the periodic table of elements. The assumption is made that elements of the same period have similar chemical properties
and behavior.

Table IV. Comparison of Molecular Weight, Boiling Point, and Melting Point
(Degrees Centigrade) of Water to Alcohols

Compound	mol wt	bp	mp
$CH_3CH_2CH_2OH$	60	97	−127
CH_3CH_2OH	46	78	−117
CH_3OH	32	65	−98
HOH	18	100	0

As mentioned previously, the total energy of hydrogen bonds in ice is about 10 kcal per mole, whereas the heat of fusion is only 1.44 kcal per mole. This suggests that 1.44/10 or about 14 percent of the hydrogen bonds in ice are broken upon melting, and therefore "normal" liquid water is about 86 percent hydrogen-bonded.

Since liquid water is a conglomerate of hydrogen-bonded pseudocrystalline icelike structures interspersed with random-length polymers, heat can be absorbed with concomitant hydrogen-bond breakage without greatly altering the chemical behavior of water, within the temperature range of living systems (i.e., random polymer lengths would be shorter). This hydrogen bonding is considered to be a "temperature buffer" or temperature-regulating mechanism in cells as transient polymers are depolymerized and re-formed. This allows cells to adjust to sudden relatively drastic temperature changes for short periods of time. Formation and disintegration have been estimated to occur at a rate approaching 10^{10} times per second. The low activation energy of hydrogen bonds makes them well suited to formation and rupture within the range of life temperatures. Viscosity of liquid water is eight times less at 100 C.

The tetrahedral arrangement of water molecules in ice results in a bond distance between oxygen atoms of 2.76 nm, as indicated in Figure 11, with free space between molecules oriented in the crystal. This results in a density of 0.9. Although the oxygen–to–oxygen bond distance in water averages 2.9 nm, in water it is not less dense than ice because the crystal lattice has broken down, allowing closer packing. That is, each water molecule in liquid water is surrounded by five to six other water molecules as compared to four in an ice crystal. Also the thermal motion of molecules increases with temperature after the crystal collapses (ice melts), which then results in a decrease in density as temperature increases.

A viscous stable water polymer called *polywater* was isolated by Deryagi and Churayev in 1968. Its existence was verified and its properties were studied by Lippincott *et al.* (1969) and Donahue (1969) in the United States. Table VI summarizes some of the reported properties of polywater, which is reported to have an average chain length of $(H_2O)_{14}$.

Table V. Comparison of Molecular Weight, Boiling Point, and Melting Point (Degrees Centigrade) of Water to Chemically Similar Molecules

Compound	mol wt	bp	mp
H_2Te	129	−4	−51
H_2Se	80	−42	−64
H_2S	34	−62	−83
H_2O	18	100	0

Table VI. Comparison of Reported Properties of Water to Polywater

Property	Water	Polywater
Freezing point	0 C	-10 to -40 C
Density at 4 C	1.0	1.01 to 1.4
Refractive index	1.33	1.44
Relative viscosity	$1 \times$	$15 \times$
O–O bond distance	7.76 nm	2.3 nm
Infrared absorption	3450 cm^{-1}	1595 cm^{-1}
	1600 cm^{-1}	1410 cm^{-1}
	1400 cm^{-1}	1360 cm^{-1}

It is yet to be determined whether "polywater" occurs as such under natural conditions dissolved in "normal" water and therefore represents a percentage of normal water, or whether water was induced to polymerize under the experimental conditions. The influence of dissolved compounds and ions on polywater has not yet been established, but such information will be very interesting and undoubtedly will have strong implications in biological systems and aquatic ecology. Davis *et al.* (1971) and Rousseau (1971) have recently argued that polywater does not exist but rather that what was observed was a trace of lactic acid. A serious controversy on this topic appears inevitable.

Sometimes water will form a partial or complete dodecahedral lattice around a hydrophobic but slightly soluble molecule such as methane (CH_4) and trap the CH_4 within the "cage." This type of hydrated molecule is called a Clathrate. Since little attraction exists between CH_4 and water, the presence of a CH_4 molecule within the "cage" affords additional strength to the dodecahedral lattice via internal repulsive forces, and this increases the ability of the lattice to resist external pressure (e.g., from mechanical pressure or heat). Such a clathrate crystal will not collapse at the normal melting point of water, and in fact ice clathrates are known to crystallize at temperatures as high as 20 C in methane gas lines (Buswell and Roudebush, 1956). Microbiological production of CH_4 at the bottom of eutrophic lakes is extensive and one can only speculate that it influences the physical properties of the water.

Incomplete water "cages" will orient around certain functional groups on biological polymers in a manner similar to that described for CH_4. Functional groups such as $-CH_3$, $-CH-(CH_3)_2$, $-CH_2CH(CH_3)_2$, $-CH_2-SH$, $-CH_2-CH_2-S-CH_3$, and $-CH_2-\phi$ can all be found on the surface of protein and other cellular macromolecules. Water oriented in this manner around cellular polymers is not free liquid but is "bound," and has crystalline properties which closely resemble those of a semisolid. Water

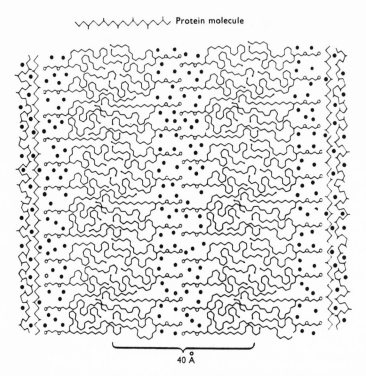

Figure 15. Double sheet cell with lipoid material alternating with protein and the whole immersed in a water medium. (From Bernal, 1966.)

bound in this manner is responsible for holding the tertiary structure (three-dimensional array or shape) of long-chain polymers in and on cells and presumably strongly influences the catalytic properties of enzymes.

It is worth pointing out in this context that all biological reactions occur at interfaces and not in bulk solution (with the possible exception of those catalyzed by extracellular enzyme reactions which are no longer cellular). Indeed, a study of cellular fine structure reveals that the cell is a series of interfolded membranous interfaces composed of a variety of chemicals which contain functional groups that are capable of forming clathrates. At the membrane level, a necessary consideration with regard to maintenance of biological organization (cellular integrity), both repulsive forces and attractive forces keep water molecules in their correct position—in and between macromolecules. Figure 15 is a schematic representation of a double unit cell membrane composed of a sandwich with lipoid material alternating with protein and the whole immersed in a water medium (Bernal, 1966).

6.2. BOUND WATER

Water molecules are oriented by all surfaces. Ionically active surfaces are always covered with layers of structured water ("ice") at normal life temperatures. This phenomenon occurs around inorganic crystals of the clay type and around all hydrophilic organic molecules such as polysaccharide and protein. For example, montmorillonite has been reported to layer water to about 40 Å thick in individual layers which are about 4 Å. The presence of this type of colloidal or amphoteric particle or large ion will produce marked disturbances in the water structure of the bulk solution and also will alter its physical properties. This highly oriented water around ions, particles, or colloidal suspensions is called *bound water* and will migrate through the bulk solution in association with whatever it is bound to. Although water bound externally on a molecule is extended in regular order, internally bound water in helical or folded molecules is not. The innermost layers of bound-water "ice layers" (e.g., around protein) are not normally penetrated by ions. These layers generally extend outward 10 to 20 Å. Beyond this to about 100 Å, water is oriented but not tightly bound, and this region will contain ions. Forces beyond this have been measured up to a 4000 Å distance into the bulk solution (Bernal, 1966). Such orientation around colloidal particles such as clay might explain how clay particles markedly increase the catalytic activity of some enzymes and why enzymes conjugated to solid surfaces in many instances become more active. It is also becoming increasingly evident to researchers that suspended colloidal particles play a dominant role in eutrophication processes in lakes. "Normal" water as we know it in the environment can therefore be considered as a loose irregular aggregation of crystalline water (ice) interspersed with linear polywater molecules which are held together by hydrogen bonds. The water molecular weight is constantly changing and can be modified by the presence of ions as well as hydrophobic and hydrophilic molecules and suspended particles— all of which would be hydrated or contain bound water.

On the biological level, water is highly organized around folded or helical membrane systems which divide cells into two different regions of water associations. One is bound (gel) and the other is "normal," in which small ions or molecules will be found.

Klotz (1962) has pointed out that the stability of many apolar hydrates of the type previously mentioned is due to the stability of the oriented water and not to any specific or unique interaction between individual apolar molecule and water. That is, most of these hydrates have hydration energies of about 16 kcal per mole. Crystalline hydrates around macromolecules including biological polymers (nucleic acids, polysaccharides, proteins, some lipids) are called *hydrotactoids* and are the counterpart of clathrates. Figure 16 is a schematic illustration of such a compound. Denaturation of these macro-molecules can be interpreted as resulting from disorganization or melting of the bound ice. This could be accomplished by heat or chemicals which have the capacity to break hydrogen bonds (e.g., urea, surfactants, certain ions). Presumably, any molecule which has a greater affinity for water (greater hydration energy) could replace the bound water, within constraints imposed by the hydrated volume of the denaturant, and therefore denature the macromolecule or cause it to precipitate from solution or colloidal suspension.

The analogy between inability of ions to penetrate the innermost layer of bound water and masking of certain functional groups on proteins has also been described by Klotz. For example, silver ion may not be able to penetrate a strongly hydrated —SH group on a protein, but if urea were also present to disorganize the crystalline lattice of bound water around the hydrotactoid, the silver ion could penetrate and react with the —SH group. Figure 17 is a schematic model for ionic hydration showing the structure of water in the neighborhood of inorganic ions. The irrotationally bound layer represents very large energies of hydration. The sphere of influence extends outward to a zone in which normal liquid water tends to form crystalline arrays—in this case shown as hexagonal arrays.

The gels that are present as capsules around many bacteria and other aquatic microorganisms result when water binds strongly to the polymers

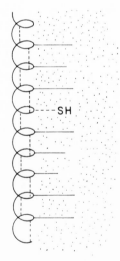

Figure 16. Schematic diagram of hydrotactoid formation around apolar groups of a protein macromolecule. (Reproduced from Klotz, 1962.)

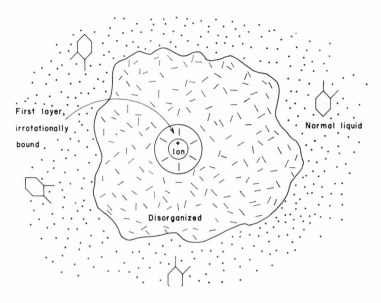

Figure 17. Model of structure of water in neighborhood of inorganic ions. (Reproduced from Klotz, 1962.)

that surround the cells. It is likely that the highly oriented water prevents penetration of certain ions to the cell surface while allowing other ions to penetrate. It is also probable that certain metal ions are adsorbed and complexed in the capsule layer by exchanging with the bound water molecules. This phenomenon has great significance in both biological flocculation processes and the reaction of colloids (microbial or other) with metal ions to cause precipitation, and biological waste treatment processes take advantage of this natural phenomenon.

6.3. BIOLOGICAL IMPLICATIONS OF BOUND WATER

The following hypothetical exercise (adopted from Snell *et al.*, 1965) serves to illustrate the quantity of hydrogen bonds involved in a cell. If we have a spherical bacterium with a diameter of $2\,\mu$, the radius would be $1\,\mu$ or 10^{-4} cm. The volume of the cell is

$$V = \tfrac{4}{3}\pi r^3 = \text{approximately } 4.2 \times 10^{-12}\,\text{cm}^3 \tag{6.1}$$

The surface area $A = 4\pi r = $ approximately 1.3×10^{-7} cm^2. Recall that the cell is about 80 percent water. Of the 20 percent solids, the bulk is protein (approximately 70 percent in terms of molecules involved). A kilogram of cells contains 800 g of water or 44.4 moles. Assuming that the average molecular weight of protein representing most of the cell solids is approximately 500, there would be $200/500 = 0.4$ mole of protein. Therefore, each protein molecule would contain or be associated with an average of $44.4/0.4 = 111$ molecules of water. Experimental data indicate that there are 960 HOH molecules per molecule of the protein gelatin when in the gel form and an average of 0.3 g of water per gram of protein in colloidal form.

These hydrated protein molecules form a tremendously large surface area in comparison to the cell volume (cytoplasm) and also comprise the approximately 2000 known enzymes which catalyze the thousands of metabolic events which occur within the small volume.

Webb (1965), in a very interesting monograph on bound water, has concluded on the basis of many experiments that about 80 percent of the physiology of a cell (and therefore of any living organism) relies on the movement of bound water but not of free water.

During experiments on aerosols directed toward elucidating specific influences of bound water, he found that cell drying or desiccation removed free water extremely rapidly but did not remove bound water and that both relative humidity and temperature altered the bound water content of a cell in an aerosol. Death rate of cells was directly related to amount of bound water remaining in the cell. Bound water amounted to about 10 percent of the total water in *Serratia marcesens*, but as expected varied with each species of cell examined (5 to 30 percent) because the chemical composition varies with age of cell within a species. Death was considered to be due to a tightening of molecular structures and was associated with small activation energies. This was presumed to result from interactions among functional groups when crystalline water around them was removed. Although many of Webb's conclusions are based on survival observations during desiccation subsequent to growth of cells, the findings presumably apply to relationships in any aquatic habitat.

Some of the physiological effects of bound water loss were found to be (1) damage or alteration of RNA and DNA, both of whose functions depend upon correct position of bound water (RNA damage results in rapid uptake of uracil—and presumably of other molecules—by the cell and "overactivity" of the remaining active RNA function); (2) dissociation of lipoprotein complexes; and (3) increased susceptibility to mutation, particularly an increased frequency of irradiation-induced (UV) mutations. The biologically active structure of a macromolecule *in vivo* appears to rely on the association it has with other molecules.

Ling (1966) has suggested that living phenomena in general are based on the ability of the protein–water–ion system to undergo cooperative changes between alternative metastable states in response to minute quantities of biologically active agents and other stimuli. Protein denaturation is a special example of such a cooperative change or transition. Disruption of a hydrogen bond in a helical region would make further disruption of a near neighbor entropically more favorable. At least eight layers of water molecules sandwiched between two polar protein surfaces are in a polarized (oriented) state. Such an oriented water space is 21.6 Å in width, and since the average chain-to-chain distance between protein chains in living cells is only 17 Å, most of the cellular water is polarized. Water in the polarized multilayered state has the property of excluding solutes such as ions and sugars. Cooperative states of protein are defined not by the protein alone but also by the solutes and solvents with which the protein associates. Ling further suggests that permease induction in bacteria arises from cooperative adsorption of sugars into cellular proteins. Such cooperative adsorption determines in an all-or-none manner the selective propensity for adsorption and accumulation of the specific solute as well as the specific secondary and tertiary structures of the protein–water–solute system.

Dissolved salts have been found to directly influence exchange processes between different kinds of water molecules bound to collagen. Part of the "bound" water molecules rotate anisotropically (NMR) about an axis parallel to the fiber direction, and this could be due to chains of water molecules which occur in the fiber direction between the collagen molecules. Hydrogen-bonding salts such as NH_4Cl increase the exchange rate of bound water molecules, and greater hydrogen-bonding capacity ($CH_3—NH_2Cl$) increases the effect. Ions such as NH_4^+ and PO_4^{2-} are stabilizers of liquid water structure but destabilizers of collagen hydration structure. In contrast, guanidinium and thiocyanate destabilize liquid water structure but have no influence on hydration. Apparently, hydrogen-bonding ions compete with water for water binding sites (Berendsen, 1966).

Although most of the studies pertaining to cell bound water have been conducted on protein macromolecules, there is no *a priori* reason that polysaccharides and polynucleic acids do not behave similarly.

There is no doubt that ions play an important role in the relationship of water binding to or around functional groups of biological macromolecules.

Robinson (1966) presents evidence to indicate that water uptake by animal cells is a colloid-osmotic phenomenon and that the driving force is the colloid-osmotic pressure of the intracellular protein which is balanced by outward transport of Na^+. If the Na^+ pump fails, extracellular fluid enters. K^+ uptake appears to be inversely related to water content of cells and is likely accumulated by a selective adsorption mechanism rather than by an energy-linked ion pump. Selective adsorption to polyelectrolyte macromolecules in the cell may be related to ion size when spacing is critically close.

In the matrix of apolar groups attached to macromolecules within a cell, water forms microcrystalline lattices whose direction depends on the spatial disposition of the apolar groups. Klotz (1962) suggests three types of conduction or transport process that might be at the molecular basis of some fundamental biochemical and physiological phenomena. The first involves conduction by an H^+ ion. H^+ can be transported by a "bucket brigade" mechanism, as depicted in Figure 18. An H^+ ion which appears on the left instantly produces an H^+ ion on the right side of the chain, which produces a very high mobility of H^+ ions in water. The second type, a similar mechanism, has been suggested for transport of electrons in water and for oxidation–reduction reactions. An example of the latter is illustrated by a hydrated ferrous ion which is hydrogen-bonded to other water molecules. An electron from Fe^{2+} should enter the oxygen valence shell if the water molecule with that oxygen would give up a hydrogen atom plus an electron to the neighboring water molecule. The relay could continue until the H radical reached an OH^- which was hydrated to another Fe^{2+}. A similar mechanism has been postulated for transfer of H^- (hydride ion) through a hydrated reducing group (S^-).

Proton conduction may also play a role in enzyme catalysis—especially hydrolysis, which may involve an acid–base mechanism. Acidic (BH) or basic (B) side chains on a macromolecule are represented in Figure 19 in proximity to an active site (X) at which a substrate molecule becomes bound. The BH or B groups could be substantially removed from the active site X and still take up or provide a proton to site X through the mediation of a water polymer.

Modification of specific acidic or basic groups would remove their potential contribution to catalysis and so would changes in apolar side chains because they are instrumental in maintaining the water bridge from B or BH to X.

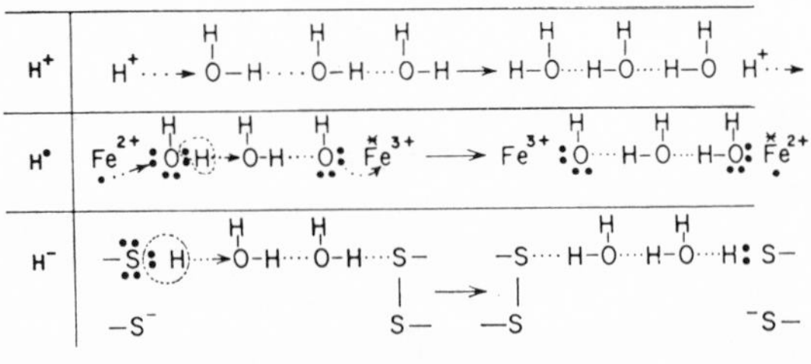

Figure 18. Drawing representing the mobility of H^+, H, and H^- through chains of water molecules. (Reproduced from Klotz, 1962.)

The solubility of gases in water varies with the chemical nature of the gas. However, those gases such as oxygen and nitrogen which do not react chemically with water have relatively low solubility in it. Gases such as carbon dioxide and ammonia react with water and have relatively high solubility in it. The solubility of all gases in water decreases with rise in temperature and increases with increased pressure.

The relationships for nonreactive gases are described by three laws: (1) Henry's law states that the mole fraction of gas in solution, at a constant temperature, is directly proportional to the pressure of the gas. The mole fraction is expressed as $n/(n + N)$, where n is the number of moles of gas and N is the number of moles of water. (2) Boyle's law states that the volume of a given gas at a constant temperature varies inversely with pressure of the gas. At a constant temperature, the product of the pressure (P) and volume (V) is a constant value, or $P_1 V_1 = P_2 V_2 = K$. (3) Charles' law states that at a constant volume the pressure of a given gas varies directly with the absolute temperature.

These relationships can be generalized as

$$pv = nRT \tag{6.2}$$

where p is the pressure, v the volume of the moles of gas, n the number of moles of gas, and T the absolute temperature. R is the molal gas constant and is the same for all gases. In the case of gases which do not react chemically with another solute or the solvent, the solubility of the gas will be less in the presence of another solute than in pure water. If the other solute reacts chemically with the gas, the solubility of the gas will be increased.

These relationships are highly significant to pollutional considerations. For example, as water becomes more polluted, i.e., more solutes are introduced, oxygen will become less soluble. As the average temperature of water increases due to such factors as use of water to cool power reactors, the solubility of oxygen will decrease. One would also expect the solubility of methane to be greater at the bottom of lakes because of the increased pressure. As will be discussed in a subsequent section, methane is produced biologically and significantly affects the carbon balance of eutrophic lakes.

As the variety of solutes in water increases, which is another way of describing pollution, more consideration must be given to solutions of liquids in water and to dispersions of immiscible liquids in water. Low molecular weight alcohols, ketones, and organic acids are polar liquids which are readily soluble in water. Many nonpolar liquids such as fats, oils, and hydrocarbons are not readily soluble in water, but the presence of surface-active solutes, known as emulsifying agents, promotes the dispersion of these compounds in water. Synthetic detergents and many proteins are effective dispersants. Many compounds of this type can also be suspended in water after being adsorbed to insoluble microscopic particulate material.

Perhaps the most familiar type of solute in water is the solid solute. During solution, the forces of aggregation in the solute are overcome by forces of dispersion exerted by the solvent phase. Everyone is aware that sugar and salt dissolve in water and go into true solution. This need not be elaborated upon here.

Insoluble substances can often be divided fine enough to be held for relatively long periods in a water column. For example, if sand is shaken with water, some will settle out immediately and some will remain suspended and the water will remain turbid. Eventually, the sand particles will settle and the water will clarify.

Many substances can be subdivided until the particles will remain suspended indefinitely. Microparticulates in this state are referred to as *colloids*. Colloids have microparticles that have physical properties between solutions and suspensions, and the colloidal state is related to the size and the chemical structure of the particle. Because chemical composition has an influence, it is not possible to set arbitrary size limits to separate solutions, suspensions, and colloids. However, many particles in the size range 0.1 μ to 1.0 mμ will result in colloid formation. Particles smaller than this range tend to dissolve, and larger particles tend to be either suspended or have no relationship to the discussion. Colloids have two phases, a dispersed, internal, or discontinuous phase and the external or continuous phase. When a colloid has a fluid appearance, it is called a *sol* and when the structure is more rigid than a fluid it is called a *gel* (e.g., jelly or gelatin gel). Sols in which the dispersed phase has little affinity for the dispersion liquid are called *suspensoids*, and those in which an affinity exists are called *emulsoids*.

The chemical and physical properties of colloids, suspensions, and solutions are of paramount importance to biology and to pollutional considerations but are much too extensive to receive adequate treatment in this monograph. It should be pointed out that many microorganisms fall in the colloidal size range, which is probably extended above the 0.1 μ size in this case because the surfaces of the organisms often have an affinity for water. Most of the high molecular weight polymers synthesized by biological systems in the aquatic habitat are also either soluble or colloidal in property.

6.5. SURFACES AND ADSORPTION

Forces exist between molecules which cause them to aggregate, and these forces can be exerted between molecules on the surface of suspended particles. The forces (e.g., van der Waals, H-bond, electrostatic) have been

Special mention should be given to the influence of sodium chloride on surfaces. The chloride ion induces a dipole moment on surface nonpolar molecules, which attract and bind the anion (Cl^-) with energy greater than the thermal or kinetic energy of the ion. The cation (Na^+) is more highly hydrated than the anion, and a high binding energy is favored by low presence of hydration water. This often results in the *indirect ionization* of nonpolar molecules in the following manner:

$$nonpolar - Cl^- + Na^+ \text{ hydrate} \qquad (6.5)$$

Both colloidal-sized microparticles including living microorganisms and dissolved chemicals adsorb to surfaces or collect at interfaces. Indeed, colloidal-sized particles are surfaces. In the case of microorganisms, the significance of the high surface-to-volume ratio has already been pointed out. In the case of enzyme molecules adsorbed to surfaces, both an increase in concentrations of enzyme and substrate and an increase in catalytic activity due to unfolding of the molecule result.

Microbial cells grow much more efficiently when adsorbed to surfaces than when in a suspended state under most natural conditions. Zobell (1943) demonstrated that numbers of bacteria increase in relationship to the container size (surface-to-volume ratio) under experimental conditions. The presence of suspended microparticles of talc, clay, plastic, sand, glass, porcelain, etc., stimulates growth of bacteria, and presumably other microorganisms, in dilute nutrient solution. This is interpreted as a surface phenomenon since there is no nutrient value to most of the microparticles studied. Many organic compounds are known to be adsorbed from solution onto glass surfaces. This is the basis of the Henrici (1933) slide technique for isolating organisms.

In summary, it can be stated that microorganisms and chemical molecules tend to accumulate at interfaces, and therefore biochemical reactions are significantly greater at interfaces than in bulk solution. Interfaces must be interpreted broadly. For example, a lake bottom: lake water is an interface on a grand scale but when examined as a series of interfaces on a microscopic scale, much greater biological significance is attached to it. Every fish, plant, microorganism, colloidal particle, and gas bubble represents an interface which results in an overall stimulation of an infinite variety of chemical reactions.

Chapter 7

DEGRADATION OF ORGANIC POLLUTANTS

Organic pollutants, as described briefly under the heading of population growth and its by-products, consist principally of animal excreta (domestic sewage) and refuse plant and animal material. Microorganisms are estimated to account for about 90 percent of the mass of fecal discharges and for this reason are a major component of domestic sewage.

Major constituents of organic pollutants on a weight basis are polysaccharides (carbohydrates), polypeptides (proteins), fats, nucleic acids, and an almost infinite variety of combinations of the above, e.g., peptidoglycans, lipopolysaccharides, phospholipids, and polyribose nucleic acids (RNA) (see Tables VII to XI).

Domestic sewage can be considered primarily as animal waste, ignoring for the moment the input of plant material from paper and kitchen disposal units, etc. Some major differences exist between plant and animal wastes.

Plant materials have a high content of the polysaccharide cellulose and lignin, a recalcitrant heteropolycyclic aromatic containing phenolic and methoxy groups. Humic acids are related to lignin and presumably contain partial degradation products. Animal wastes are relatively higher in protein and lipid materials and therefore in sulfur and nitrogen as protein components.

The main means of degrading biological polymers to their monomeric units is via hydrolytic reactions. Hydrolysis by definition involves the incorporation of water into the molecules as the polymer is broken down. The enzymes which catalyze the reactions are known collectively as *hydrolases*. Individual hydrolases are quite specific for individual substrates and are predominantly extracellular or cell-surface enzymes.

Tables VII to XI present a chemical summary of human excretory products which comprise a major portion of domestic sewage (Albritton, 1952). The average mineral and organic content of domestic sewage is shown

Table VII—Continued

Constituent	Excreted in urine (mg/kg body wt/day)		Excreted in feces (mg/kg body wt/day)		Excreted in sweat (mg/100 ml)	
	Value	Range[b]	Value	Range[b]	Value	Range[b]
Threonine						
Free	0.37	0.17–0.62				
Combined	0.4	0.3–0.8				
Total	0.77	0.36–1.2	4.0	3.5–5.2	5.4	1.7–9.1
Tryptophan						
Free	0.37	0.12–0.7				
Combined	0.3	0.009–0.4				
Total	0.7	0.23–1.3			1.1	0.4–1.8
Tyrosine						
Free	0.3	0.17–0.55				
Combined	0.5	0.08–0.9				
Total	0.79	0.35–1.45			3.2	1.2–5.0
Valine						
Free	0.065	0.04–0.125				
Combined	0.2	0.09–0.4				
Total	0.3	0.21–0.45	4.6	3.6–6.2	3	1.5–4.5
Miscellaneous compounds						
Methione sulfoxide		0–0.31				
Indican	0.14	0.06–0.45				
Adrenaline[h]	0.16	0.07–0.31				
Taurine		0.105–0.2				
Allantoin	0.27	0.18–0.36				
Noradrenaline[h]	0.4	0.17–0.9				
Purine bases	0.41	0.18–0.92		2–3		
Guanidoacetic acid		0.23–0.51				
Histamine		0.2–1.0				
Creatine[i]	2.9	1.1–3.86				
Hydroxytyramine[h]		1.4–2.8				
Imidazole derivatives		1.35–9.4		0–0.2		

*From Albritton, 1952.

[a] Because of the high degree of variability in rate of sweat formation, ranging from zero under some conditions up to as high as 12 liters per day in extremely hot climates, it has not been practicable to present data on excretion via sweat in terms of "per kg body weight per day." (From Albritton, 1952.)

[b] Ranges are averages of ranges of values reported in the literature cited.

[c] Nitrogen in excreta is present as nitrogen compounds and not as free nitrogen.

[d] Total N, and NPN values have been calculated from the values listed for the individual nitrogen components, items No. 6–11. See also footnote b.

[e] See also Table XI.

[f] Determined by paper chromatography, identity not completely proven.

[g] Identity not proven.

[h] The catecol amines are expressed in micrograms.

[i] Not normally present in the urine of adult males.

in Table XII on the basis of grams per capita per day and also on the basis of milligrams per liter for an average sewage flow rate of 100 g per capita per day. The values are broken down into the categories of dissolved solids and suspended solids, which in turn are subdivided into settlable (large particulate) and nonsettlable (colloidal dispersion) solids.

As mentioned previously, most waste treatment methods in use today are biological processes which take advantage of the catalytic and metabolic activities of microorganisms for the purpose of converting pollutants to more highly oxidized and hence less oxygen-demanding forms. The primary constituents of all organic wastes can be grouped together as polysaccharides, fats, and proteins. The general biochemical sequences which organisms

Table VIII. Excretion of Lipids by Man*

Constituent	Excreted in urine (mg(kg body wt/day) Value	Range[b]	Excreted in feces (mg/kg body wt/day) Value	Range[b]	Excreted in sweat (mg/100 ml) Value	Range[b]	Excreted in sebum[a] (g/100 g) Value
Fat							
Total			56	30.0–100			
Neutral				10–45			
Unsaponifiable			33	22–38[c]			
Fatty acids							
Total			30	4–64			
Free			16	4–38			28
Unsaturated							15[d]
Cholesterol							
Total	0–0.007		8[e]	10–20[e]			5[d]
Free							2.5[d]
Paraffin							7.5[d]
Phosphatides							0.96[d]
Soaps			53[c]	40–66			
Squalene							5[d]
Triglycerides							33[4]
Waxes[g]							15[d]

*From Albritton, 1952.
[a]The data on sebum are not available in mg per kg body weight per day, and are therefore presented in g per 100 g.
[b]Ranges are averages of ranges of values reported in the literature cited.
[c]Age 8–12 years.
[d]From forearms only.
[e]Age 10 months.
[f]Individual samples from forehead.
[g]Include esters of cholesterol.

Table XI. Excretion of Electrolytes and Minor Minerals by Man[*,a]

Constituent	Excreted in urine (per kg body wt/day) Value	Range[b]	Excreted in feces (per kg body wt/day) Value	Range[b]	Excreted in sweat (per 100 ml) Value	Range[b]
Aluminum, μg		0.7–1.6	0.6			
Ammonia,[c] μg	9,200	4,900–18,200		360–1,200		2,500–35,000
Arsenic, μg	0.46	0–1.15	33	1–116		
Bromine, μg		12–110				
Calcium, μg	2,900	1,100–4,910	7,490	5,000–10,000	2,060	100–5,500
Chlorine,[d] mg	115	84–193		0.21–0.5		30–300
Cobalt, μg	0.07	0.05–0.12	0.007	0.002–0.02		
Copper, μg	2.38	0–7.52	27	23–37	6.0	
Fluorine,[e] μg		6.7–100[f]				
Iodine,[g] μg	1.4	0.2–2.13			0.8	0.5–1.2
Iron, μg	0.7	0.7–1.4	120	65–208	27	22–45
Lead, μg	0.5	0.06–2.1	4.2	2.2–19.8		
Magnesium, μg	1,850	950–4,500	2,500	1,510–3,185	200	140–4,500
Manganese, μg		0.095–1.4		18–120	6	3–7
Mercury, μg		0.007–0.01	0.14			
Nickel, μg	2.1	2–4		1.2–2.5		
Nitrates, μg	7,140					
Phosphorus, mg	15	10–19	9.86	7.1–20	1.5	0–4.8
Potassium, mg	34	14–46	6.7			21–126
Selenium, μg	1	0–3.3				
Silicon, μg	108	14–200				
Silver, μg			0.8			
Sodium, mg	46	38–91	1.7			29–294
Sulfur						
Total, mg	16.5	4–20.6	2.0			0.7–7.4
Ethereal, mg	1	0.6–4.3				
Inorganic, mg	12.5	3.5–18.25				
Neutral, mg	1.9	1.0–3.0				
Tin, μg		0.13–0.31		170–450		
Zinc, μg	4.6	1–6.4	101	46–500		

*From Albritton, 1952.
[a]Because of the high degree of variability in rate of sweat formation, ranging from zero under some conditions up to as high as 12 liters per day in extremely hot climates, it has not been practicable to present data on excretion via sweat in terms of "per kg body weight per day." (From Albritton, 1952.)
[b]Ranges are averages of ranges of values reported in the literature cited.
[c]See also Table VII.
[d]Chloride.
[e]Fluoride.
[f]Data include regions in Texas where dental fluorosis is endemic.
[g]Iodide.

It should be born in mind that all reactions outlined in Figure 20 are those which are exploited in biological waste treatment processes. Specific

Table XII. Average Domestic Sewage*

	Mineral	Organic	Total	5-day BOD
	grams/capita/day[a]			
Suspended solids	25	65	90	42
Settlable	15	39	54	19
Nonsettlable	10	26	36	23
Dissolved solids	80	80	160	12
Total solids	105	145	250	54
	in mg/liter for a flow of 100 gal/capita/day[b]			
Suspended solids	65	170	235	110
Settlable	40	100	140	50
Nonsettlable	25	70	95	60
Dissolved solids	210	210	420	30
Total solids	275	380	655	140

*From Imhoff and Fair, 1966.
[a]1 g/capita = 2.2 lb/1000 population.
[b]1 mg/liter in a per capita volume of n gallons of sewage daily = 0.00378 n g/capita daily.

wastes will vary in percent composition of the major constituents depending upon origin and may therefore require special handling as an engineering consideration. For example, fruit cannery wastes would undoubtedly have a high percentage of carbohydrate, whereas abattoir wastes would be high in protein. Specific examples of the general reactions presented in Figure 20 will be considered throughout this chapter.

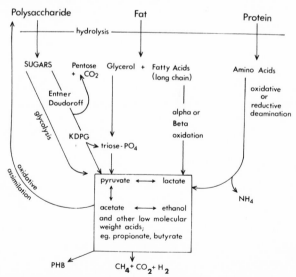

Figure 20. Flow diagram illustrating some of the major pathways for degradation of major categories of organic compounds.

occurs through alpha 1–6 linkages. Animal glycogen is very similar to amylopectin.

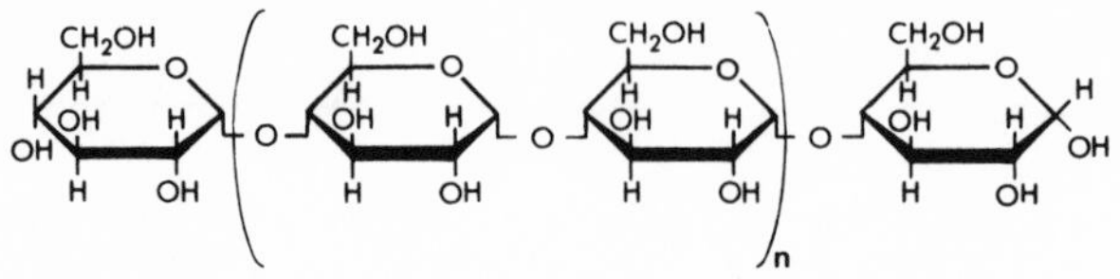

amylose
(I)

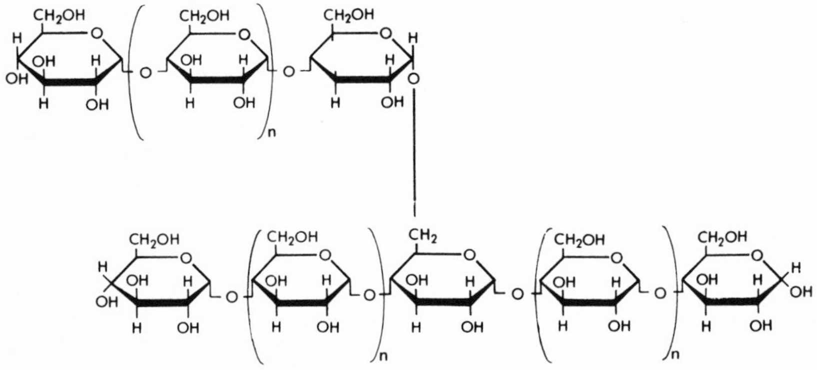

amylopectin and glycogen
(II)

Pectins (III) are polyuronic acids and are the intercellular cementing substance of plant tissue and therefore very commonly encountered as waste substance. A variety of pectins are known. They are predominantly polymers of galacturonic acid, arabinose, and methylated galacturonic acid. Many microorganisms produce pectin-decomposing enzymes, which are familiar to most as "soft rot" of vegetables and fruits. The enzyme degradation of pectin has long been utilized in the retting of flax fibers to make linen as well as to make mats from coconut fibers.

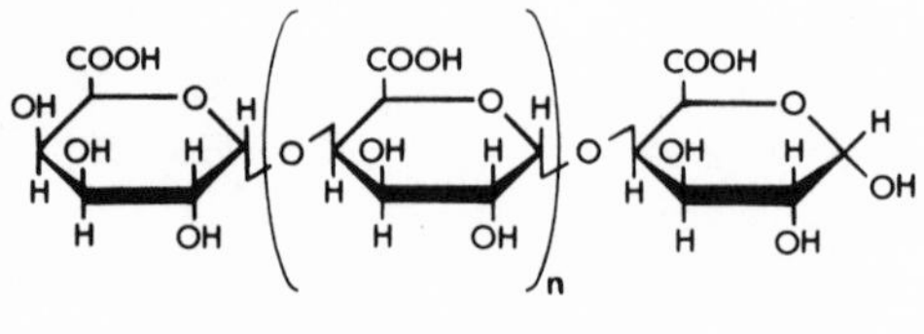

pectin (polyuronides)
(III)

Another common group of plant polysaccharides is called *hemicellulose*. The most common hemicelluloses are xylans (IV), which are polyxylose with an occasional arabinose unit and/or methylglucuronic acid. Since xylose and arabinose are five-carbon sugars, the xylans are also called *pentosans*. They are found in woody plant tissue and also comprise about 50 percent of the dry weight of straw.

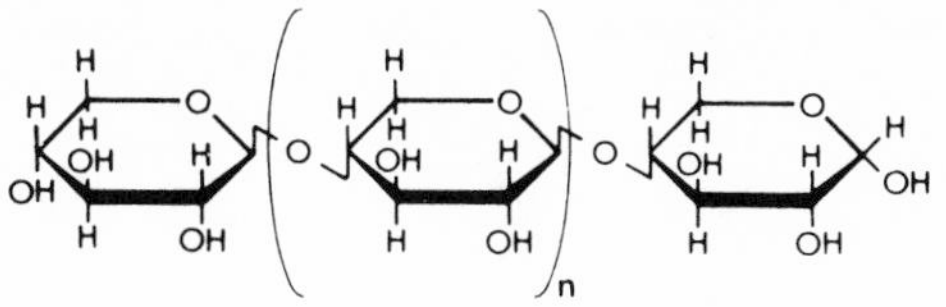

xylan
(IV)

Cellulose (V) is one of nature's most common polysaccharides. It is polyglucose linked by beta 1–4 bonds. Cellulose is more resistant than many other polysaccharides, primarily because relatively few organisms produce enzymes that are able to attack a beta 1–4 linkage. The hydrolytic reaction also proceeds at a slower rate than many other hydrolytic reactions, presumably because the polymer has a higher molecular weight than many others (200,000 to 1,000,000), which makes it insoluble. All biological reactions proceed in aqueous surroundings, and water-insoluble molecules are attacked biologically at the water-substrate interface.

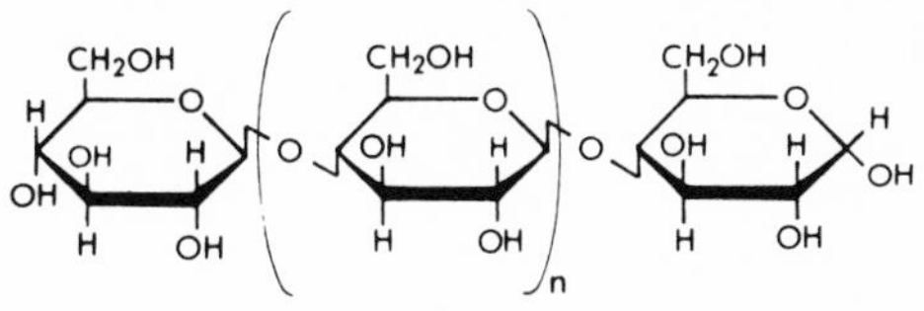

cellulose
(V)

Polysaccharides are commonly synthesized by bacteria, algae, and other microorganisms and are much more prevalent and significant in the aquatic environment than is usually considered. Some of the implications are discussed in Chapter 12.

There are many other naturally occurring polysaccharides and polysaccharide derivatives. Natural plant gums (e.g., gum acacia, gum arabic, gum ghatti) are polysaccharides. Agar and algin are polysaccharides that contain galactose-6-sulfate. *Chitin*, the resistant substance found in crayfish exoskeletons and human fingernails is poly-*N*-acetyl glucosamine, a polysaccharide derivative.

Hydrolytic enzymes are produced by both aerobic and anaerobic organisms, and the reactions do not require the presence of oxygen *per se*.

7.4. CARBOHYDRATE DISSIMILATION

Sugars are degraded via fermentation or oxidative pathways. Disaccharides such as sucrose and lactose are hydrolyzed to monosaccharides in a manner similar to that described for polysaccharides. Monosaccharides are metabolized by conversion to glucose or fructose or their respective phosphate esters and then degraded fermentatively to pyruvic acid via the Embden–Meyerhof (glycolytic) pathway, as outlined in the following sequence of reactions:

$$
\begin{array}{c}
\text{glucose} \longrightarrow \text{fructose diphosphate} \\[4pt]
2\,\text{ATP} \quad 2\,\text{ADP} \\[4pt]
2\ \text{phosphoglyceric acid} \longleftarrow 2\ \text{triose phosphate} \\[4pt]
\longrightarrow 2\ \text{pyruvic acid}
\end{array}
\tag{7.3}
$$

Pyruvic acid occupies a unique central role in carbon metabolism in that it can be directly converted to lactic acid, acetic acid, amino acid (alanine), oxalacetic acid, acetaldehyde, and then to ethanol, or it can be oxidized to carbon dioxide and water via the tricarboxylic acid cycle (TCA cycle). Acetic acid can be condensed to butyric and several other fatty, hydroxy, or keto acids as well as to the polymer-β-hydroxybutyric acid (PHB). PHB is a naturally occurring carbon storage product synthesized by many bacteria. It has a regular helical structure consisting of repeating units of β-hydroxybutyric acid (Alper $et\ al.$, 1963). Data have been presented which suggest that PHB in all microorganisms is synthesized from the same basic molecule but that the molecular weight ranges from 1000 to 250,000 (Lundgren $et\ al.$, 1965). PHB is a polyester in which the carboxyl group of each hydroxybutyrate is esterified with the hydroxyl group of another molecule in the following way:

$$
\underset{\substack{\\ \text{CH}_3\ \ \text{H}\ \ \text{O}}}{\text{HO}-\text{C}-\text{C}-\text{C}}\left(\underset{\substack{\\ \text{CH}_3\ \ \text{H}\ \ \text{O}}}{\text{O}-\text{C}-\text{C}-\text{C}}\right)_n \underset{\substack{\\ \text{CH}_3\ \ \text{H}\ \ \text{O}}}{\text{O}-\text{C}-\text{C}-\text{C}-\text{OH}}
\tag{7.4}
$$

A large amount of organic carbon can be assimilated rapidly into microbial cells by conversion to PHB. The process converts organic acid to a neutral polyester and therefore is somewhat analogous to a buffer system. PHB granules often nearly fill the cell volume, which may result in rupture of the cell membrane and cell lysis. PHB may be depolymerized enzymically during periods of low carbon nutrition, or the fat-soluble granules may remain intact as cells lyse. (For general considerations of the above pathways in addition to other reactions in this chapter, the reader is referred to Sokatch, 1969; Stanier *et al.*, 1963; Thimann, 1968.)

In addition to oxidation of hexoses by the TCA cycle, they can be oxidized by the oxidative pentose cycle and by the Entner–Doudoroff pathway as shown in (7.5). These latter two pathways are characteristic of *Acetobacter*, *Acetomonas*, and *Pseudomonas* bacteria, all of which are indigenous to aquatic habitats.

Oxidative pentose shunt

Entner–Doudoroff pathway

(7.5)

Such reactions can be generalized as

$$4 H_2A + CO_2 \rightarrow 4 A + CH_4 + 2 H_2O \tag{7.7}$$

where A is a primary or secondary alcohol.

Acetic acid can be cleaved via a fermentation to methane and carbon dioxide:

$$CH_3{-}COOH \rightarrow CH_4 + CO_2 \tag{7.8}$$

The fermentation of propionic acid is more complex:

$$4CH_3{-}CH_2{-}COOH + 8 H_2O \rightarrow 4CH_3COOH + 4 CO_2 + 24 H \tag{7.9}$$

$$3 CO_2 + 24 H \rightarrow 3 CH_3 + 6 H_2O \tag{7.10}$$

$$Sum: 4CH_2CH_2COOH + 2 H_2O \rightarrow 4 CH_3COOH + CO_2 + 3 CH_4 \tag{7.11}$$

During the methane fermentation, the oxygen-consuming equivalent of organic matter removed is

$$CH_4 + 2 O_2 \rightarrow CO_2 + 2 H_2O \tag{7.12}$$

The molecular weights are 16 and 64; therefore, 16 g of CH_4 produced is equivalent to removal of 64 g of oxygen (BOD removal) in streams.

Propionibacterium is an interesting gram-positive microaerophilic catalase-positive rod, and therefore not an anaerobe, which will, however, not grow exposed to air. This bacterium converts pyruvic acid to propionic acid plus acetic acid:

$$3 CH_3{-}\overset{\overset{\displaystyle O}{\|}}{C}{-}COOH + H_2O \rightarrow CH_3CH_2COOH + 2 CH_3COOH + 2 CO_2 \tag{7.13}$$

7.5. AMINO ACID DISSIMILATION

Amino acids can be biologically dissimilated by any of several enzymic reactions, as shown below using alanine as an example. The type of reaction depends upon the particular enzymes elucidated by the particular organism, and this in turn depends to a large extent on environmental conditions such as presence or absence of oxygen. Some of the reactions are summarized in the following scheme. The reactions are referred to as *oxidative*, *reductive*, or *hydrolytic deaminations* where the amine group is split off to form NH_3. The reactions in which carbon dioxide is split off are referred to as *decarboxylations* [see (7.14)].

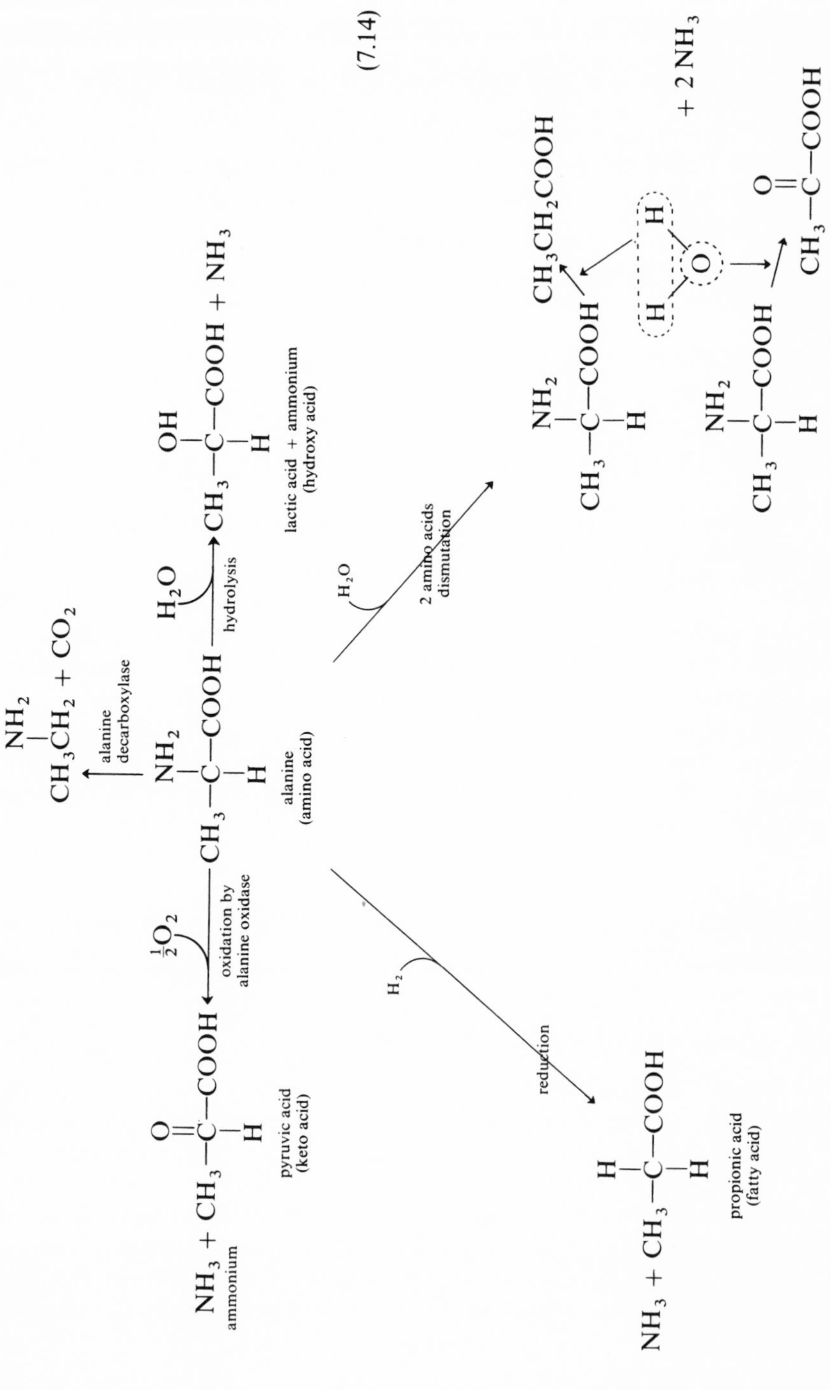

$$
\begin{array}{c}
NH_2 \\
|\\
CH_3CH_2 + CO_2
\end{array}
$$

alanine decarboxylase

$$NH_3 + CH_3\!-\!\overset{O}{\overset{\|}{C}}\!-\!COOH$$
$$|$$
$$H$$

pyruvic acid (keto acid)

ammonium

$\tfrac{1}{2}O_2$ oxidation by alanine oxidase

$$CH_3\!-\!\overset{NH_2}{\overset{|}{C}}\!-\!COOH$$
$$|$$
$$H$$

alanine (amino acid)

H_2O hydrolysis

$$CH_3\!-\!\overset{OH}{\overset{|}{C}}\!-\!COOH + NH_3$$
$$|$$
$$H$$

lactic acid + ammonium (hydroxy acid)

(7.14)

H_2O 2 amino acids dismutation

$$CH_3\!-\!\overset{NH_2}{\overset{|}{C}}\!-\!COOH \qquad CH_3CH_2COOH$$
$$|$$
$$H$$

$$CH_3\!-\!\overset{NH_2}{\overset{|}{C}}\!-\!COOH \qquad CH_3\!-\!\overset{O}{\overset{\|}{C}}\!-\!COOH$$
$$|$$
$$H$$

$+\ 2\ NH_3$

H_2 reduction

$$NH_3 + CH_3\!-\!\overset{H}{\overset{|}{C}}\!-\!COOH$$
$$|$$
$$H$$

propionic acid (fatty acid)

$$H_2N-CH_2-CH_2-CH_2-\overset{\overset{\displaystyle NH_2}{|}}{C}H-COOH \xrightarrow[\text{decarboxylase}]{\text{ornithine}}$$

ornithine

$$CO_2 + H_2N-CH_2-CH_2-CH_2-CH_2-NH_2 \quad (7.24)$$

putrescine

$$H_2N-CH_2-CH_2-CH_2-\overset{\overset{\displaystyle NH_2}{|}}{C}H-COOH \xrightarrow[\text{decarboxylase}]{\text{lysine}}$$

lysine

$$CO_2 + H_2N-CH_2-CH_2-CH_2-CH_2-CH_2-NH_2 \quad (7.25)$$

cadaverine

Hydrogen sulfide is produced enzymically from sulfur-containing amino acids by some bacteria and is of value in differentiating species of bacteria which produce the enzymes responsible for release of H_2S into the surrounding medium. One example of such a reaction is

$$HS-CH_2-\overset{\overset{\displaystyle NH_2}{|}}{C}H-COOH + H_2O \xrightarrow[\text{desulfhydrase}]{\text{cysteine}}$$

cysteine

$$H_2S + NH_3 + CH_3\overset{\overset{\displaystyle O}{||}}{C}-COOH \quad (7.26)$$

pyruvic acid

One additional reaction of considerable significance indirectly related to amino acid degradation is the decomposition of urea by the enzyme unrease (7.27). This enzyme is produced by both aerobic and anaerobic organisms.

$$NH_2-\overset{\overset{\displaystyle O}{||}}{C}-NH_2 + H_2O \xrightarrow{\text{urease}} CO_2 + 2\,NH_3 \quad (7.27)$$

7.6. OXIDATION OF ALCOHOLS AND FATTY ACIDS

Once oxygen has been incorporated into a hydrocarbon or a hydro-carbon derivative, it becomes an alcohol, aldehyde, fatty acid, or ketone. Alcohols, aldehydes, fatty acids, and ketones can also be formed biologically in several ways—for example, by transamination or deamination of amino acids, hydrolysis of fats, fermentation of sugars, and by photosynthetic and chemoautotrophic pathways. Once formed, they are attacked biologically in a variety of ways, and the origin of the compound is irrelevant.

Alcohols are usually converted via a two-step process to their corresponding fatty aldehyde (7.28) and then to the acid (7.29). This is illustrated by the conversion of ethanol to acetaldehyde and then to acetic acid by two different dehydrogenase enzymes which are coupled to reduction of the coenzyme NAD. Longer-chain alcohols would be oxidized to corresponding long-chain acids, and the acids would be subsequently oxidized via beta oxidation.

$$CH_3CH_2OH \xrightarrow{\text{alcohol dehydrogenase}} CH_3CHO \qquad (7.28)$$
$$NAD \quad NADH \cdot H$$

$$CH_3CHO \cdot H_2O \xrightarrow{\text{acetaldehyde dehydrogenase}} CH_3COOH \qquad (7.29)$$
$$NAD \quad NADH \cdot H$$

Alpha hydroxy fatty acids can be attacked by a mechanism referred to as *alpha oxidation* (7.30), which requires a two-step reduction and an oxidative decarboxylation:

$$RC \overset{O}{\underset{OH}{\diagup}} \quad + CO_2 + H_2O \qquad (7.30)$$

7.6.1. Beta Oxidation

Long-chain fatty acids are attacked via a sequence of reactions called *Beta oxidation*, in which two carbon units at a time are split from the chain. The sequence (7.31 to 7.35) is shown for removal of an acetate unit from a pentanoic acid, ultimately liberating propionic acid plus acetate, which is combined with reduced coenzyme A (CoASH). Acetic acid can either be liberated from CoA in a subsequent reaction or the acetate can be transferred to other functional compounds in a cell.

Chapter 8

HYDROCARBON OXIDATION

8.1. GENERAL ASPECTS

Hydrocarbons are very ubiquitous in nature. Crude petroleum is primarily a hydrocarbon mixture which is divided into various fractions that can be marketed. Gasoline is the fraction that consists primarily of hexane (C_6), heptane (C_7), and octane (C_8) and the carbon skeletons can be in a straight chain, branched, or cyclic. Other fractions, used for other purposes, consist of mixtures of higher or lower molecular weight hydrocarbons, e.g., natural gas, kerosene, motor oil. The hydrocarbon molecules can also be either saturated or unsaturated.

Unsaturated hydrocarbons are very common in biological systems. Several types are synthesized by plant cells, and waste products of vegetable origin therefore contain rather significant amounts of hydrocarbons and related compounds, e.g., carotenoids, natural rubber (polyisoprene), vitamins K_1 and K_2, coenzyme Q. The incidence of water pollution by crude oil has stimulated a renewed interest in this area.

In this regard, Zobell (1946) stated that all kinds of gaseous, liquid, and solid hydrocarbons of the aliphatic, olefinic, aromatic, or naphthenic series appear to be susceptible to oxidation by microorganisms, provided that the hydrocarbons are properly dispersed. He also suggested that aliphatic hydrocarbons were oxidized more readily than aromatic or naphthenic compounds. Within certain limits, long-chain hydrocarbons are attacked more readily than similar compounds of low molecular weight. It also appears that the addition of aliphatic side chains increases the susceptibility of cyclic compounds to microbial attack. Among the hydrocarbons occurring in the aquatic environment, in addition to petroleum products, and therefore subject to biological oxidation are synthetic detergents, pesticides, and other biochemically important compounds. Zobell stated that the disappearance of oil from waterways, from soil around refineries, from leaking pipelines, and from polluted beaches is due largely to the activity of hydrocarbon-oxidizing microorganisms. Since most petroleum products are paraffinic in character, their mode of oxidation would follow that of the alkanes and iso-alkanes. McKenna and Kallio (1964) stated previously that increased branching of alkyl chains in general reduces the availability of these compounds to microbial attack. From these statements it would seem that gasoline, having multiple branching, would be very resistant to microbial oxidation, and this was in fact reported by Zobell.

With reference to hydrocarbon oxidation or degradation as a pollutional consideration, it may be of some concern whether the hydrocarbons are used for growth by the organism or whether the hydrocarbons are co-oxidized in the presence of another nutrient. The ecological reasons for concern are (1) in the case of growth, one would anticipate an enrichment and selection of populations which could use the hydrocarbons, and (2) in the case of co-oxidation, one would anticipate by-product production which might then accumulate in the environment and cause secondary interactions.

A large proportion of bacterial species and other organisms can oxidize saturated fatty acids and fatty alcohols, but fewer species are able to oxidize the corresponding alkanes. This implies that once oxygen has been incorporated into a saturated alkane hydrocarbon molecule it is metabolized with relative ease by a variety of organisms and that the key reaction in alkane oxidation is the initial incorporation of oxygen. Although few species oxidize hydrocarbons, this still represents a significant variety of microorganisms. McKenna and Kallio (1964) showed that from 10 to 20 percent of the groups of bacteria, yeasts, and fungi examined were able to degrade alkanes.

It is known that bacteria oxidize the terminal methyl group of straight-chain hydrocarbons to corresponding alcohols and acids by the incorporation of oxygen directly into the molecule. For example, Baptist *et al.* (1963) were able to show that octanol, octaldehyde, and octanoic acid were produced from octane by cell-free extracts of *Pseudomonas oleovorans*.

The mechanism of enzymic incorporation of oxygen into a saturated alkane hydrocarbon is not well understood but can be inferred by comparison to the oxidation of related compounds.

Two classes of enzyme are known to incorporate molecular oxygen directly into organic molecules. Enzymes in the first class are known as *oxygenases* or *oxygen transferases*, and those in the second class are called *hydroxylases*. Oxygenases typically catalyze the introduction of both atoms of bimolecular O_2 into their specific substrate according to the following scheme:

$$RH_2 + O_2 \xrightarrow{\text{oxygenase}} R(OH)_2 \tag{8.1}$$

Three general oxygenase types are known.

Oxygenase type (1) consists of those that cleave aromatic rings. Examples of this type of reaction are

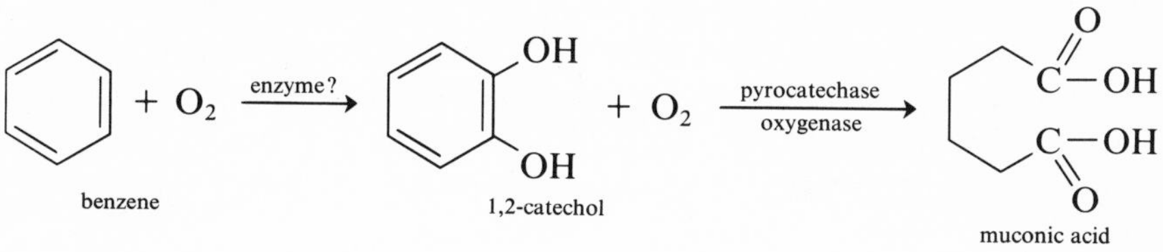

(8.2)

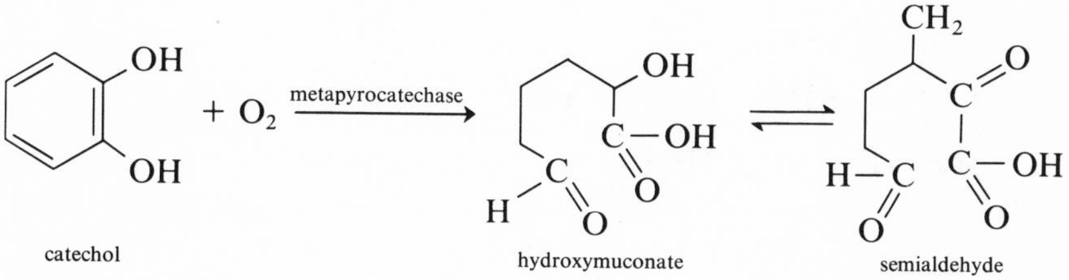

(8.3)

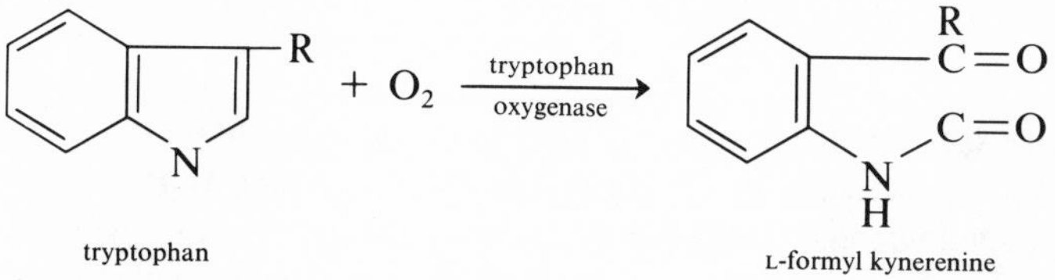

(8.4)

These enzymes are present in some species of the bacterium *Pseudomonas* and require iron cofactors.

Oxygenase type (2) is those that oxygenate nonaromatic rings. The rings are cleaved but re-form with one less carbon atom. Reaction (8.5) is an example of this type, which has been observed in kidney cells but has not been observed in microorganisms.

Oxygenase type (3) is those that oxidize lipids which contain CIŚ-CIŚ-1-6 pentadiene to the corresponding hydroperoxide. These enzymes are found in plant cells and are known as *lipoxygenases* or *lipoxidases* and catalyze the reaction shown in (8.6). Reactions of this type have been observed in plant cells.

Hydroxylases, members of the second class of enzymes which incorporate molecular oxygen directly into substrate, have also been called *mixed-function oxygenases and mono-oxygenases* and are different from oxygenases in that they characteristically catalyze incorporation of a single atom of O_2 directly into the substrate.

$$(8.5)$$

$$(8.6)$$

Most studies on the mechanism of enzymic reactions involving both oxygenases and hydroxylases have utilized substrates other than hydrocarbons (e.g., *Pseudomonas*, lysine hydroxylase; *Streptomyces*, arginine oxygenase; *Mycobacterium*, lactate oxygenase). Those that have been studied require NADH, Fe^{2+}, and O_2 and often require a second enzyme, presumably to remove the initial reaction product (e.g., octane → octanol would require an alcohol dehydrogenase to further oxidize the alcohol). Once a straight-chain alcohol has been oxidized to its corresponding acid, it can be degraded via beta oxidation or sometimes via omega oxidation to dicarboxylic acids.

Saturated hydrocarbons have also been observed to yield corresponding ketones (Leadbetter and Foster, 1959; Lukins and Foster, 1963) when attacked by *Pseudomonas methanica* and *Mycobacterium* species, in which case the oxygen is incorporated onto the alpha carbon rather than the terminal carbon.

8.2. *n*-ALKANES

Microbial attack on alkanes primarily occurs at the terminal carbon atom and results in the early formation of a fatty acid corresponding to the chain length of the alkane (Kallio *et al.*, 1963). Oxygen can be introduced into the alkane molecule in at least two ways. The first (8.7) is hydroperoxidation of the alkane with subsequent formation of an alkyl hydroperoxide (Stewart *et al.*, 1959). This is decomposed to a primary alcohol, which is then oxidized to the primary acid. Further attack occurs by the beta oxidation of fatty acids.

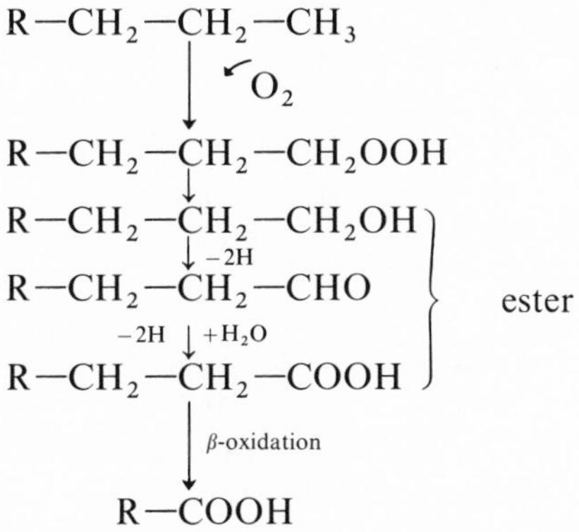

$$R-CH_2-CH_2-CH_3$$

$$\downarrow \nwarrow O_2$$

$$R-CH_2-CH_2-CH_2OOH$$

$$\left. \begin{array}{l} R-CH_2-CH_2-CH_2OH \\ \quad\downarrow_{-2H} \\ R-CH_2-CH_2-CHO \\ \quad_{-2H}\downarrow_{+H_2O} \\ R-CH_2-CH_2-COOH \end{array} \right\} \text{ ester} \qquad (8.7)$$

$$\downarrow \beta\text{-oxidation}$$

$$R-COOH$$

In the second pathway (8.8), dehydrogenation of the alkane to an olefin occurs, followed by the addition of water to the double bond to give the primary alcohol.

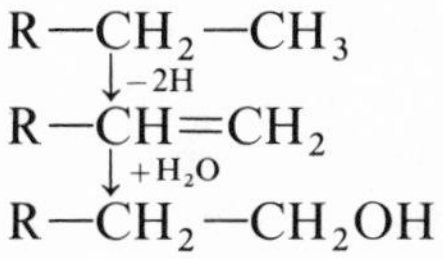

$$R-CH_2-CH_3$$
$$\downarrow_{-2H}$$
$$R-CH=CH_2 \qquad (8.8)$$
$$\downarrow_{+H_2O}$$
$$R-CH_2-CH_2OH$$

Leadbetter and Foster (1960) also observed the formation of methyl ketones in the fermentations of short-chain *n*-alkanes. They proposed a free-radical equilibrium between primary and secondary carbons (8.9). Hydroperoxidation could occur at either the primary or secondary carbon.

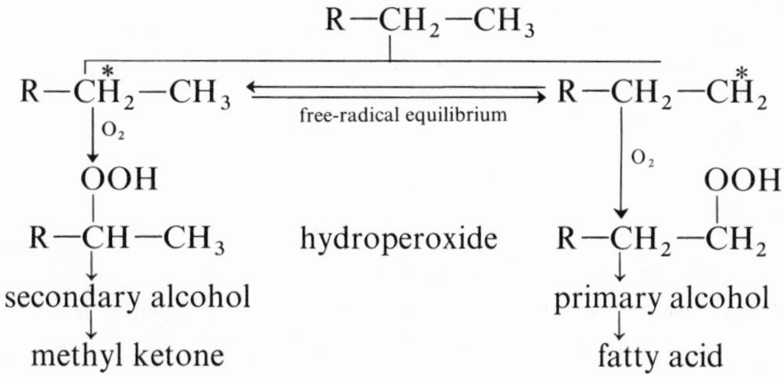

$$R-CH_2-CH_3$$

$$R-\overset{*}{C}H_2-CH_3 \underset{\text{free-radical equilibrium}}{\rightleftharpoons} R-CH_2-\overset{*}{C}H_2$$

$$\downarrow O_2 \qquad\qquad\qquad \downarrow O_2$$

$$OOH \qquad\qquad\qquad OOH$$
$$| \qquad\qquad\qquad\qquad | \qquad (8.9)$$
$$R-CH-CH_3 \quad \text{hydroperoxide} \quad R-CH_2-CH_2$$
$$\downarrow \qquad\qquad\qquad\qquad \downarrow$$
$$\text{secondary alcohol} \qquad\qquad \text{primary alcohol}$$
$$\downarrow \qquad\qquad\qquad\qquad \downarrow$$
$$\text{methyl ketone} \qquad\qquad\qquad \text{fatty acid}$$

Kester and Foster (1963) observed a diterminal oxidation of *n*-alkanes (8.10). The sequence involved seems to be monoterminal oxidation, probably by hydroperoxidation, followed by omega oxidation. Further breakdown occurs by beta oxidation at either end of the molecule.

$$CH_3-(CH_2)_x-CH_3$$
$$CH_3-(CH_2)_x-COOH$$

$$\text{omega} \Big| \text{or} \searrow \beta\text{-oxidation} \tag{8.10}$$

$$HOOC-(CH_2)_x-COOH$$
$$\beta\text{-oxidation}$$

Leadbetter and Foster (1960) grew *Pseudomonas methanica* on a growth medium that contained methane as a carbon source. Although the organism was unable to grow at the expense of other alkanes, it was capable of oxidizing them in the presence of methane, a process referred to as *co-oxidation*. In this case, ethane, propane, butane, and hexane were co-oxidized to the corresponding alcohols and fatty acids.

8.3. ALKENES

Several metabolic pathways have been suggested for the oxidation of alkene-1 compounds (8.11). Ishikura and Foster (1961) and Stewart *et al.* (1960) suggested the primary pathway to be by way of oxidation of the terminal methyl group at the saturated end of the molecule. They also believed that some microorganisms may also initiate attack on the unsaturated end of the molecule, but that it may be a minor pathway. Again, subsequent carbon chain breakdown presumably involves beta-oxidation.

main pathway   minor pathways

$$\overset{OH\quad OH}{CH_3(CH_2)_x CH=CH_2 \xrightarrow{Candida\ lipolytica} CH_3(CH_2)_x CH-CH_2}$$

$$\downarrow \substack{Pseudomonas \\ aeruginosa} \qquad\qquad \substack{Micrococcus \\ cerificans}$$

$$HOCH_2(CH_2)_x CH=CH_2 \xrightarrow{} CH_3(CH_2)_x CHCOOH \tag{8.11}$$

$$\downarrow$$

$$HOOC(CH_2)_x CH=CH_2$$

8.4. CYCLOALKANES

Until a few years ago, the cycloalkanes were thought to be resistant to microbial oxidation. Ooyama and Foster (1965) reported on the bacterial oxidation of cycloparaffinic hydrocarbons. They suggested that once the pristine hydrocarbon molecule is breached, the biochemistry ceases to be distinctly hydrocarbon and metabolism is determined by the particular carbon skeleton of the substrate. They indicated that saturated and unsaturated cyclic compounds are attacked by dihydroxylation of vicinal ring carbons, yielding the corresponding *trans*-1:2-diol:

$$\rightarrow \text{degradation} \qquad (8.12)$$

These same investigators also described the formation of cyclic mono-ketones (8.13) from cyclic hydrocarbons. Present evidence indicates that the corresponding alcohol is the precursor of the cycloketone and that oxygenation of the hydrocarbon ring takes place in the molecular oxygen, probably via hydroxylase. The alcohol would then be dehydrogenated to the ketone.

cyclopentane   cyclopentanone   cyclohexane   cyclohexanone

cyclohexanol   cyclohexanone   cyclohexene   cyclohexanone

methylcyclohexane   4-methylcyclohexanone

$$(8.13)$$

8.5. AROMATICS

The general mechanism of ring splitting of aromatic hydrocarbons is shown in (8.14) taken from Gibson *et al.* (1967) and Humphrey (1967). The ring is oxidized by perhydroxylation and then dehydrogenated to the dehydroxy compound (Evans, 1963). Ring cleavage takes place by oxidative fission of the bond between the carbon atom bearing the hydroxy groups forming *cis-cis* muconic acid (Evans *et al.*, 1951). This is further oxidized in β-oxadipic acid (Kibley, 1948; Stanier *et al.*, 1950), which is enzymatically split to acetyl-CoA and succinate and shunted into the TCA cycle. An alternative pathway occurs via the alpha keto acids.

Humphrey (1967) reported that three key intermediates exist in the scheme of single-ring aromatic oxidations: catechol, protocatechuic acid, and genitistic acid (8.15). This sequence indicates how various aromatic hydrocarbons are broken down via various oxidative pathways.

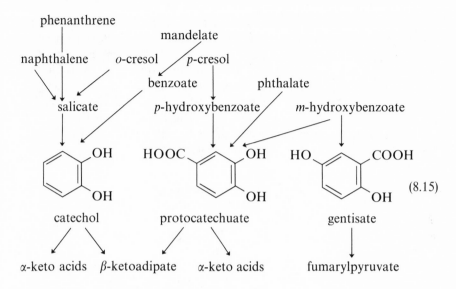

$$(8.15)$$

8.6. METHANE, ETHANE, AND METHANOL OXIDATION

Methane gas (CH_4) deserves special consideration because it is present in copious amounts in water columns (Dugan *et al.*, 1970*b*). Methane is produced biologically under anaerobic conditions from organic compounds that accumulate at the water–bottom interface of eutrophic lakes, ponds, swamps, etc. It is also produced by the same biological process during anaerobic digestion of sewage. Although methane has limited solubility in water, it has the capacity to form water clathrates, and its continuous biological production makes it continuously available to oxidative attack as it rises from the bottom into aerobic zones of the water column. Lack of methane accumulation in nature suggests its continuous removal.

The autotrophic bacterium *Methanomonas* has been shown to oxidize CH_4 by the following general reaction:

$$CH_4 + 2O_2 \rightarrow CO_2 + 2H_2O + \text{energy} \qquad (8.16)$$

Leadbetter and Foster (1960) isolated pigmented pseudomonads (*Pseudomonas methanica*) that both oxidized CH_4 to CO_2 and incorporated CH_4 into cell substance.

Ethane-oxidizing bacteria (mycobacteria) have also been reported (Dworkin and Foster, 1956).

Several pathways have been suggested for bacterial oxidation of methane and have been reviewed by Humphrey (1967). It is generally agreed that the first step is the incorporation of oxygen into methane to form methanol (8.17). The subsequent steps of the process are not clear. Harrington and Kallio (1960) suggested a peroxidative attack of the methanol to form formaldehyde followed by dehydrogenation to formic acid. Dworkin and Foster (1956) suggested dehydrogenation of methanol to formaldehyde, oxidation to formic acid, and finally oxidation to give carbon dioxide. Another method proposed by Johnson and Quayle (1965) suggests methane assimilation at reduction levels between methanol and formate. Reduced C_1 units are incorporated into sugar phosphates.

$$
\begin{array}{c}
\underset{\text{form}}{\text{reduced}} \;\; \overset{?}{\dashrightarrow} \; \text{sugar phosphate} \\[4pt]
\nearrow^{\;?} \\
CH_4 \xrightarrow{\;O_2\;} CH_3OH \xrightarrow{\;H_2O_2\;} CH_2O \xrightarrow[H_2O]{\;-2H\;} HCOOH \xrightarrow{\;?\;} CO_2 \qquad (8.17) \\
\underset{-2H}{\searrow} \\
CH_2O \dashrightarrow^{\;?} HCOOH \xrightarrow{\;-2H\;} CO_2
\end{array}
$$

Some species of acetic acid bacteria are capable of oxidizing methanol to formic acid or to carbon dioxide and water (Asai, 1968) by reactions which are referred to as *oxidative fermentations*. Methanol utilization by *Zoogloea* strains has been reported by Friedman and Dugan (1968) and Joyce and Dugan (1970). *Zoogloea* has been reported to be closely related to *Acetobacter* on the basis of immunological similarities (Schmidt *et al.*, 1970), and the lack of either acid formation or carbon dioxide production from methanol may be due either to ester synthesis or to incorporation into cell substance (Joyce and Dugan, 1969).

Chapter 9

RECALCITRANT MOLECULES

9.1. ISOALKANES

Isoalkanes are branched hydrocarbons and are more resistant to enzymic degradation than are straight-chain hydrocarbons, and branched chains are more likely to be encountered, on a chance basis, with longer chain length or higher molecular weight. Unsaturated hydrocarbons are more amenable to degradation than saturates. The oxidation pathways of the iso-alkanes have not been fully elaborated, although considerable work has been done. Humphrey (1967), however, made several generalizations: (1) straight-chain alkanes are more easily oxidized than brached chains, (2) compounds containing one methyl branch are attacked only when the molecule contains a sufficiently long unbranched chain, and (3) alkanes with a branched alkyl group larger than methyl or with multiple methyl branches are not oxidized. A few other generalizations with regard to branching have been summarized by McKenna and Kallio (1964) on the basis of microbial growth on hydro-carbon substrates. Addition of a second CH_3 group into a saturated straight chain usually makes the molecule more recalcitrant to a variety of genera and species of organisms. That is, a quaternary carbon atom makes the compound more refractory to enzymes, particularly if it is adjacent to the terminal carbon atom. This is schematically diagrammed below (9.1), where carbon skeletons of hydrocarbon molecules are shown and molecules are presented in decreasing order of susceptibility to enzyme attack:

$$(C-C-C-C-C)_n > (C-C-C-\underset{\underset{C}{|}}{C}-C-C-C-\underset{\underset{C}{|}}{C}-C)_n \tag{9.1}$$

$$> (C-C-C-\underset{\underset{C}{|}}{C}-C-C-C-\underset{\underset{C}{|}}{C}-C-C)_n \quad > C-\underset{\underset{C}{|}}{\overset{\overset{C}{|}}{C}}-C-\underset{\underset{C}{|}}{\overset{\overset{C}{|}}{C}}-C-\underset{\underset{C}{|}}{\overset{\overset{C}{|}}{C}}-C$$

Since a straight chain can be oxidized from either or both ends, branching or blockage of each end of the molecule must be considered. In the case where a benzene (phenyl) ring is added to an alkane molecule, we have the following order of decreasing susceptibility:

$$\phi\,C-C-C-C-C-C-C-C-C-C$$

$$> \phi\,C-C-\underset{\underset{C}{|}}{C}-C-C-C-C-C-C-C$$

$$> \phi\,C-C-C-\underset{\underset{C}{|}}{\overset{\overset{C}{|}}{C}}-C-C-C-C-C-C$$

$$(9.2)$$

Location of the phenyl group also influences susceptibility in the following decreasing order:

$$\underset{\underset{\phi}{|}}{C}-C-C-C-C-C-C-C-C > C-\underset{\underset{\phi}{|}}{C}-C-C-C-C-C-C-C$$

$$> C-C-\underset{\underset{\phi}{|}}{C}-C-C-C-C-C-C$$

$$(9.3)$$

High molecular weight polymers can be synthesized from hydrocarbon monomers (e.g., polyethylene, polypropylene, butyl rubber; see Table XVI) which are quite resistant to decomposition and constitute a major source of environmental clutter.

Experimental data have been obtained by Nickerson (1969) which indicate that many of the synthetic high polymers (e.g., polyethylene, rubber) can be biologically degraded via inducible enzymes provided the induction process is properly maintained.

9.2. SYNTHETIC ANIONIC DETERGENTS

Extrapolation of the above generalizations to environmental significance must be made with caution. If hydrocarbon-degrading activity of a few species having high population densities as the result of enrichment more than compensates for lack of species diversity, the net effect will not be diminished. An environmental scale observation relevant to this point is the case of "hard" vs. "soft" detergents. Some anionic detergents were referred to as "hard" because they resisted enzyme attack. The detergent alkylbenzene sulfonate (ABS) is a chemical derivative of the hydrocarbons in (9.2) and (9.3).

alkylbenzene sulfonate (ABS)
(VII)

linear alkylbenzene sulfonate (LAS)
(VIII)

It consists of a benzene ring with an alkane side chain terminating in a hydrophilic sulfonic acid group (VII). The alkane side chain in commercial preparations of ABS consisted of all possible branched arrangements of carbon atoms with C_{12} being the predominant size. After several years of environmental contamination because of concentration increase due to enzyme refractility, the soap and detergent industry substituted a linear side chain (LAS) in place of the branched chains (VIII). The detergent pollution problem has abated considerably because the linear LAS molecule is degraded by microbial enzymes in the environment, and is referred to as "soft" (Sawyer and Ryeknan, 1958; Simko et al., 1965). Also, several field tests have indicated that LAS appears to be as biologically degradable as other soluble naturally occurring organic compounds found in water (Renn et al., 1964; Swisher, 1963). Huddleston and Allred (1963) suggested that beta oxidation is the mechanism of LAS degradation.

Widespread accumulation of ABS in the environment between 1945 and 1965 gives some insight into our previous question and suggests that fewer species will not proliferate sufficiently to compensate for lack of enzyme susceptibility. There is some consensus among researchers that enzymes do not attack branched chains readily because of steric hindrance of the branched molecule. The sulfonic acid group is subject to enzyme hydrolysis by a variety of aquatic bacteria and is not responsible for the recalcitrant nature of the molecule.

Detergents are highly surface-active chemicals and are effective in low concentrations. Commerically available proprietary grade detergents usually contain less than 30 percent by weight active detergent. The remainder of the material in the supermarket package consists of "filler," water softener, fluorescent whitening agents, etc. One of the major problems related to detergent use is the high proportion of phosphate chemicals that are added as water-softening agents. Several chemical forms of phosphate are in use, but all react with Ca^{2+}, Mg^{2+}, and Fe^{3+} ions—the primary ions causing hardness in water, thereby softening the water.

Phosphates ultimately find their way into receiving waters, where they are utilized as nutrients by algae and other organisms. Stimulated algal growth is a problem discussed under eutrophication (Chapter 12).

Compounds such as borate and aminonitrilotriacetate have been contemplated for use as substitutes for phosphate. It is difficult to envision a net gain from an antipollutional viewpoint if this is carried out. Nitrogen, boron, and acetate all can stimulate algal growth. Borate can also be toxic to animals in relatively low concentrations.

9.3. HYDROCARBON DERIVATIVES

Hydrocarbons can be made more recalcitrant by substituting halogens on the molecules. Halogenation increases the molecular resistance to degradation in the following order of decreasing resistance:

$$F > Cl > Br \text{ or } I$$

Polymerization and cross-linkages also increase resistance of the molecule to enzyme or chemical attack. Examples of resistant hydrocarbons and hydrocarbon derivatives of the type that accumulate as environmental clutter are shown in Table XVI.

Chlorinated aromatic hydrocarbons have long been utilized as antibiological agents. Chlorinated phenols have been used extensively as industrial germicides for control of microbial slime formation and other nuisance bacteria. The compound sodium pentachlorophenate (IX) is marketed for this use and the compound G-11 hexachlorophene (X) has been added to soap, shaving cream, etc., as a topical germicide. Although there are few data available, it is interesting to speculate on the effects of the trend toward chlorination of sewage. Chlorine adds very readily to phenols and unsaturated hydrocarbons to form chlorinated hydrocarbons, and reference to Table VII shows that a significant amount of phenols and unsaturated

(IX)   (X)

Table XVI. Structures of Some Recalcitrant Hydrocarbon Derivatives That Accumulate in the Environment

Monomer	Polymer	Product
H H | | C=C | | H H Ethylene	H H H H H H H | | | | | | | —C—C—C—C—C—C—C— | | | | | | | H H H H H H H	Polyethylene
H H | | C=C | | H Cl Vinyl chloride	H Cl H Cl H Cl H | | | | | | | —C—C—C—C—C—C—C— | | | | | | | H Cl H Cl H Cl H	Polyvinyl chloride
H Cl | | C=C | | H Cl Vinylidine chloride	H Cl H Cl H Cl H | | | | | | | —C—C—C—C—C—C—C— | | | | | | | H Cl H Cl H Cl H	Polyvinylidine chloride (Saran)
	H H H H H H H | | | | | | | —C—C—C—C—C—C—C— | | | | | | | OH HOH HOH HOH	Polyvinyl alcohol (water soluble)
H CH$_3$ | | C=C | | H CH$_3$ Isobutylene	H CH$_3$ H CH$_3$ H CH$_3$ H | | | | | | | —C—C—C—C—C—C—C— | | | | | | | H CH$_3$ H CH$_3$ H CH$_3$ H	Polyisobutylene (component of butyl rubber)
F F | | C=C | | F F Tetrafluoroethylene	F F F F F F F | | | | | | | —C—C—C—C—C—C—C— | | | | | | | F F F F F F F	Polytetrafluroethylene (Teflon)

organics are present in the excreta of humans and presumably that of other animals as well. There is cause for suspicion that we are unwittingly adding to our pollution problems as the result of our attempts to avoid them. On the other hand, chlorination of sewage is an attempt to smooth over a distressing situation but not an attack on the source of the problem.

9.3.1. CHLORINATED HYDROCARBON PESTICIDES

Relatively few of the approximately 900 active pesticidal chemicals formulated into over 600,000 proprietary preparations are chlorinated hydrocarbons. Although most pesticides are potentially hazardous environmental contaminants, the chlorinated hydrocarbon group presents the greatest hazard because it contains chemicals that are highly toxic to most biological systems and they are also highly resistant to degradation. The total 1967 production of pesticides of all types in the United States was about 5×10^5 tons, of which more than 50 percent was organochlorine types. Surveys indicate that use of insecticides will double by 1975 and herbicide use will increase to more than twice that of insecticides during the same period. On a world basis, the production would be 25 to 50 percent greater than the United States production.

The significance of recalcitrance of highly toxic compounds, hence accumulation in the environment, in addition to direct toxicity lies in the

Table XVII. Some Metabolic Reactions Common to Pesticides*

Pathway	Schematic	Pesticide which undergoes this reaction
1. Oxidation		
a. Hydroxylation		2,4-D
b. Side-chain oxidation		DDT
c. Ether cleavage		2,4-D
d. Sulfoxide formation	$R-S-CH_3 \rightarrow R-\overset{O}{\underset{\mid}{S}}-CH_3$	Phorate
e. N-oxide formation		Schraden

Table XVII —Continued

Pathway	Schematic	Pesticide which undergoes this reaction
2. Dehydrogenation and dehydrohalogenation		DDT
3. Reduction	$R-NO_2 \rightarrow RNH_2$	DNOC
4. Conjugation		
a. Amide formation	$R-NH_2 + R'-COOH \rightarrow R-\overset{H}{\underset{\vert}{N}}-\overset{O}{\overset{\parallel}{C}}-R$	Amitrole
b. Metal complex	$R_2-N + MeX \rightarrow (R_2N-Me)X$	Amitrole
c. Glucoside and glucuronic acid		Barthrin
d. Sulfate	$ROH + SO_4 = \rightarrow R-O-\overset{O}{\underset{\underset{O}{\parallel}}{\overset{\parallel}{S}}}-O^-$	Biphenyl
5. Hydrolytic		
a. Cleavage of esters	$R-\overset{O}{\overset{\parallel}{C}}-OR' \rightarrow R-\overset{O}{\overset{\parallel}{C}}-OH + HOR'$	Malathion
b. Cleavage of amides	$R-\overset{O}{\overset{\parallel}{C}}-\overset{H}{\underset{\vert}{N}}-R' \rightarrow R-\overset{O}{\overset{\parallel}{C}}-OH + R'-NH_2$	Dimethoate
6. Exchange reactions	$R-O-\overset{S}{\overset{\parallel}{P}}-(OR)_2 \rightarrow R-O-\overset{O}{\overset{\parallel}{P}}-(OR)_2$	Parathion
7. Isomerization	$R-O-\overset{S}{\overset{\parallel}{P}}-(OR)_2 \rightarrow R-S-\overset{O}{\overset{\parallel}{P}}-(OR)_2$	Most organophosphates

* From Menzie, 1969.

insidious potential for sublethal chronic effects, heritable alterations (mutagenesis), effects on reproduction (teratogenesis), and cancer formation (carcinogenesis).

High-risk, persistent chlorinated hydrocarbons include dichloro-diphenyl trichloroethane (DDT), dichlorodiphenyldichloroethane (DDD), aldrin, dieldrin, isodrin, endrin, heptachlor, chlordane, and benzene hexachloride (BHC), including lindane, the gamma isomer of BHC. The elimination of all uses of DDT and DDD within 2 years and drastically restricted use of the others has been recommended (Report of the Secretary's Commission on Pesticides, 1969). The available literature on pesticides is quite voluminous, and many reviews on specific interest topics concerning pesticides are also available. The general types of metabolic reactions which are known to involve all types of pesticide substrates have been summarized by Menzie (1969) in the following manner:

1. Oxidation (in the sense that oxygen, as hydroxyl, takes part or is postulated to take part in one or more of the steps)
    (a) Hydroxylation of aromatic rings
    (b) Oxidation of side chains to alcohols, ketones, or carboxyl groups
    (c) Dealkylation from oxygen or sulfur (ether cleavage)
    (d) Sulfoxide formation
    (e) N-oxide formation

2. Dehydrogenation and dehydrohalogenation

3. Reduction

4. Conjugation
    (a) Amide formation
    (b) Metal complex
    (c) Glucoside or glucuronic acid
    (d) Sulfate

5. Hydrolytic reactions
    (a) Cleavage of esters
    (b) Cleavage of amides

6. Exchange reactions

7. Isomerization

Each of these types of reaction is illustrated in Table XVII for a pesticide known to undergo the particular reaction. It is evident that the general reactions of the chlorinated hydrocarbons are essentially the same as for hydrocarbons. Specific known biochemical reactions of some of the most recalcitrant and problem-causing chlorinated pesticides are shown in Figures 22 through 26, taken from Menzie (1969).

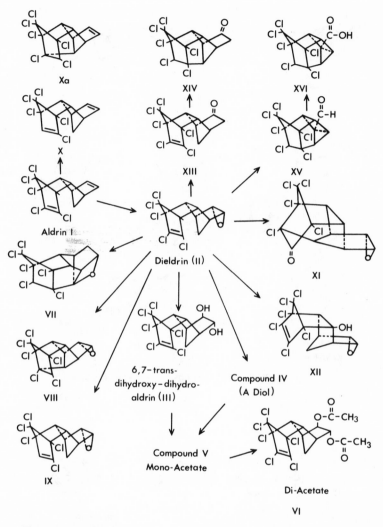

Figure 22. Biochemical conversion of aldrin and dieldrin. (From Menzie, 1969.)

Figure 22 shows chemical conversions of aldrin and dieldrin. Figure 23 illustrates reactions of isodrin and endrin. Figures 24 and 25 show reactions of DDT, DDD, and intermediate reaction products. Figure 26 presents reactions of benzene hexachloride (BHC, lindane).

These illustrations indicate that the compounds undergo rather minor chemical modifications in the environment, and in many cases the products still have a marked toxity, e.g., conversion of aldrin to dieldrin (Figure 22). It is

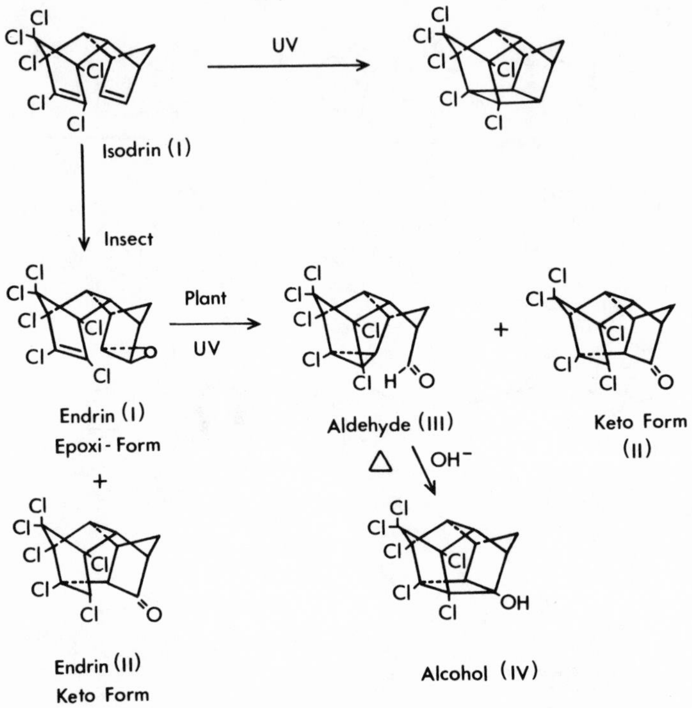

Figure 23. Biochemical reactions of isodrin and endrin.

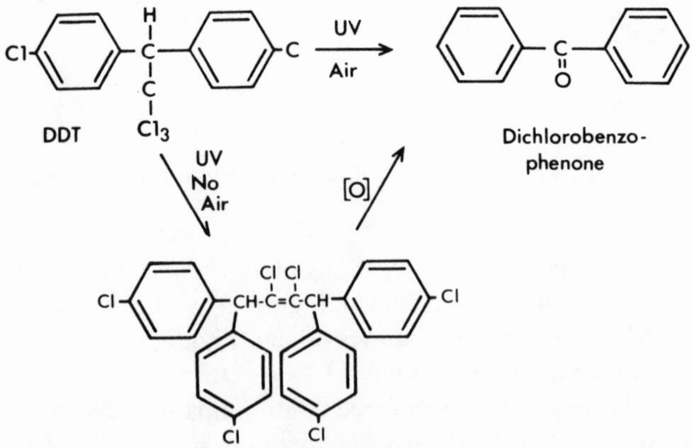

Figure 24. Biochemical reactions of DDT. From Menzie, 1969.

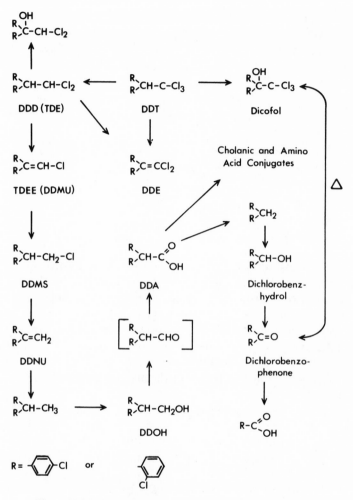

Figure 25. Biochemical reactions of DDE. ₍From Menzie, 1969.

implicit that not all cells have the capacity to catalyze all of the steps indicated by the arrows. Usually, a concerted effort by a group of organisms is required to accomplish a two- or three-step sequence.

One of the most interesting problems related to pesticide toxicity is how such nonreactive molecules can exert such toxic effects on cells. A plausible explanation has been offered by Okey and Bogan (1965). The effect of chlorine (halogens in general) on metabolism depends on the overall make-up of the molecule; i.e., alterations in the chemical reactivity of substituted organics are related to alterations in electron density at specific sites on a

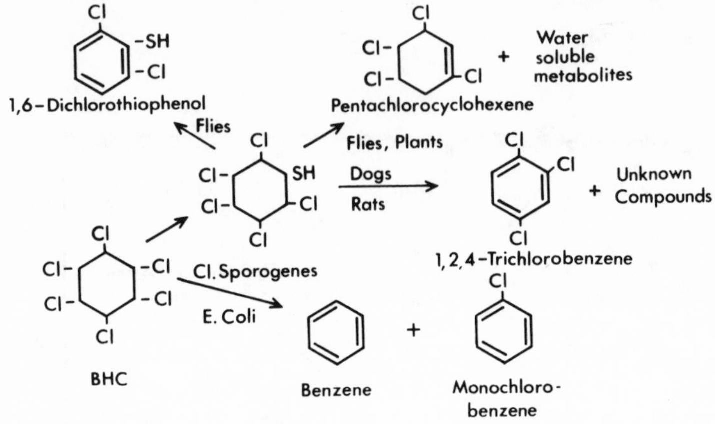

Figure 26. Biochemical reactions of benzene hexachloride. From Menzie, 1969.

substrate. The density alteration which is created by electronic and resonant properties of the substituent (Cl) alters the acceptability of the site for an attacking functional group (on an enzyme). Chloro groups are strongly electronegative and attract electrons from the remainder of the molecule toward itself (inductive effect). Hydrogens also alter the resonant properties of aromatic or conjugated substituent groups (resonant interaction or mesomeric effect). This attraction in electron density acts to either increase or reduce the activation energy barrier through which a molecule must pass during a reaction.

Biological oxidations are electrophilic in nature, and substituent groups which lower the electron density at a reaction site will decrease the reaction rate. Enzyme catalysis is carried out by lowering the level of energy necessary in the substrate for the reaction. Electron manipulation is a requirement for catalytic activity.

Consider the reaction

$$E \text{ (enzyme)} + S \text{ (substrate)} \rightarrow [E \cdot S] \text{ complex} \xrightarrow{k} E + \text{products} \quad (9.1)$$

$E \cdot S$ is an activated complex, and k is the rate of dissociation of the activated complex and is expressed by the following equation:

$$k = \frac{RT}{Nh} \exp \frac{-\Delta F}{Rt} \quad (9.2)$$

where h is Plank's constant, ΔF the free energy of activation, R the gas constant, T the absolute temperature, and N is Avogadro's number.

The equation is not affected by reaction order or molecularity and is consistent with the electron influence postulated for Cl substituent groups on pesticide molecules. It is assumed that the steric configuration of the toxic molecule effects the ability to gain proximity to an active site on normal substrate or enzyme molecules.

Hill and McCarty (1967) demonstrated that DDT, aldrin, DDD, lindane, and endrin were degraded more rapidly under anaerobic than under aerobic conditions. Other investigators (Johnson et al., 1967; Kallman and Andrews, 1963; Stenerson, 1965; Wedemeyer, 1966) have shown that a number of different facultatively anaerobic bacteria and yeasts can dechlorinate DDT to DDD under anaerobic conditions. Dechlorination was either incomplete or occurred at a slower rate under aerobic conditions.

9.3.2. LIGNIN

Lignin is a very common but complex polycyclic aromatic compound synthesized by plants. For example, pine needles contain approximately 27 percent lignin. Although lignins undoubtedly vary markedly in chemical composition among species of plants, they are comprised of a series of cross-linked phenyl propane monomers with some phenolic groups and some methoxy groups attached to the phenyl rings. The exact chemical structure has not been determined.

Lignin sulfonate is a major by-product produced by the pulp and paper industry. It is produced as the result of separation of cellulose fibers from the lignin fraction of wood through the use of the sulfite pulping process. Lignin sulfonate forms a salt in alkaline solution and has a marked solubility as sodium lignin sulfonate. Lignin sulfonate has relatively little commercial value, and tremendous amounts are stored in large piles at pulp mills. It has a brownish-black color and represents a source of color pollution when the soluble material either leaches from storage piles into streams or is added directly to streams. Sodium lignin sulfonates are relatively more biodegradable than native lignin.

By comparison to most biologically synthesized compounds, lignin is highly resistant to biodegradation. Two groups of microorganisms, the white wood rot fungi (various species of *Basidiomycete*) and certain aerobic bacteria, are known to attack lignin oxidatively at slow rates via phenolase enzymes. *White rot* is a name given the group of fungi which appear to decompose the lignin portion of wood somewhat preferentially over the cellulose portion; thereby leaving the decaying wood "white." Bacteria that have been most often reported as lignin decomposers are *Pseudomonas*, *Flavobacterium*, and *Achromobacter* species, all of which are obligate aerobes that are common inhabitants of soil and water (Tabak et al., 1959; Sorenson, 1962). The subject of biological decomposition of lignin has been reviewed by Lawson and Still (1957).

Because lignin degradation proceeds biologically via oxidative cleavage (see Chapter 8) and not by hydrolytic reaction, it is resistant to anaerobic decomposition. This implies that as vegetable material decomposes in water, some lignin degrades as long as oxygen is present. As the material settles to the bottom of a lake or pond where anaerobic conditions often prevail, the cellulose portion will be decomposed via hydrolytic and fermentative reactions but the lignin portion will accumulate. The process contributes to build-up of organics in sediments and to the formation of bogs and ultimately to coal formation (Francis, 1954). Lignin build-up must therefore play a significant role in the aging of lakes, and the same process is responsible for production of humus in soils.

Chapter 10

CYCLING OF NUTRIENTS

The cycling of nutrient elements is the crux of all ecological considerations because of their limited availability in the biosphere. The concept of growth-limiting nutrients has been discussed in Chapter 2 as well as the essential elements for all biological systems. It follows that continued utilization of one element as a nutrient by a given category of living organism would eventually tie up all of the available supply and that no further growth could occur once the supply was depleted. For example, if all of the carbon dioxide available to photosynthetic organisms was fixed (reduced) into an organic polymer form as the result of photosynthetic activity, no further photosynthesis could take place. The time estimated for such an occurrence on earth is a matter of days. If it were not for the continued renewal of the supply of carbon dioxide by the respiratory activities of other types of organisms, all life would stop soon after photosynthesis stopped and all of the available carbon would be fixed in the form of dead carcasses, leaves, wood, etc., to be preserved forever.

The same argument holds true for each biologically essential element, and each of these elemental cycles could be discussed separately with various degrees of complexity. Not all elements are likely to become limiting nutrients for a given species because of differing concentrations of available elements in the biosphere and also because certain elements are required in much greater amounts than others by cells. It is the balance of supply and demand for the elements carbon, nitrogen, sulfur, oxygen, and phosphorus that appears to regulate the overall balance of nature. However, specific mineral requirements often regulate the extent of growth of individual species in a given environmental niche. This latter situation has great pollutional significance. For example, if molybdenum in a certain concentration is required for nitrogen fixation by nitrogen-fixing blue-green algae and it happens to be absent in a particular lake, there will be relatively little nitrogen enrichment in the lake. However, if molybdenum is incorporated as a fertilizer component and added to farm land around the lake, some may eventually be washed into the lake as a result of storm runoff and thus stimulate nitrogen fixation (enrichment) in the lake. The same considerations hold for each of the nutritional elements in one ecological situation or another.

Discussion of elemental cycles is a matter of convenience and can be misleading unless it is clearly understood that all macronutrient elements are cycled simultaneously and that the cycles cannot be divorced from one another except as a means of explanation. The sulfur-containing amino acid cysteine is often described as containing a reduced organic form of sulfur during consideration of the sulfur cycle, but inspection of the molecule, $HS-CH_2-CH(NH_2)COOH$, reminds us that it also contains carbon, nitrogen, oxygen, and hydrogen, which are components of other elemental cycles. Biological conversion of sulfur in cysteine represents an electrochemical redistribution for all atoms of the molecule. Another aspect which is often misleading is lack of appreciation for the variety, diversity, and complexity of organisms involved in completing a turn of an elemental cycle.

10.1. CYCLING OF OXYGEN, HYDROGEN, AND PHOSPHORUS

Of the macronutrient elements that are essential for all biological systems in relatively high amounts, oxygen is considered to cycle between an oxidized form (O_2) and reduced forms $(CO_2, NO_3, H_2O$, etc.). Oxygen is more appropriately considered in terms of availability balance in a water column and as an oxidizing agent than in terms of cycling. It reacts chemically with carbon, nitrogen, and sulfur and thereby effects the cycling of these elements. The same can be said of hydrogen, which cycles between H_2 and H_2O. Both oxygen and hydrogen are of prime importance in considerations of oxidation–reduction potentials in the environment but need not be further discussed under the cycling heading.

Phosphorus has virtually no biological involvement in any form other than phosphate. Cycling of phosphate is between the free ionic form and the organically bound form and therefore has a somewhat different connotation than cycling of carbon, nitrogen, or sulfur.

10.2. CYCLING OF CARBON

Carbon is the most prevalent biological element, and carbon-containing compounds are synonymous with the word "organic". It always is found in biological systems in the ± 4 oxidation state. It is the element most often chosen to describe elemental cycling. This is usually accomplished in terms of balance among the oxidative processes of repiration or combustion and the reductive processes which fix carbon dioxide into an organic form.

Photosynthetic reactions all are coupled to carbon dioxide fixation, and the fixation of carbon dioxide by chemoautotrophic bacteria has been mentioned previously.

Many heterotrophic organisms also fix significant amounts of carbon dioxide while deriving energy from the partial degradation of other organic nutrients. The mold *Neurospora crassa* can condense carbon dioxide with ornithine (Chapter 7) to form citrulline; animal cells can condense carbon dioxide with acetyl coenzyme A to form malonyl-CoA; *Aerobacter aerogenes* fixes carbon dioxide into purines and pyrimidines; and many bacteria require carbon dioxide for growth. For a review of this topic, see Thimann (1968).

Much of this monograph has already been devoted to the significance of organic carbon conversions in water pollution, and this need not be reiterated here.

10.3. CYCLING OF NITROGEN

The nitrogen cycle is illustrated in Figure 27 on the basis of the oxidation state of nitrogen in the various compounds known to be metabolized by organisms or acted upon by enzyme systems. The cycle is comprised of a series of reactions that are involved with reduction of nitrate, the most highly oxidized chemical form of nitrogen, to ammonium, the most highly reduced form of nitrogen. Ammonium can be combined biochemically to form amino acids, which can then be converted biologically to a variety of other organic nitrogen-containing compounds or back to ammonium. Release of ammonium from organic forms of all types is referred to as *ammonification*. Many of these reactions have been described in Chapter 7. Ammonification is shown as a separate but vastly important loop in the cycle because the nitrogen always stays in the same oxidation state (-3). Ammonium can also be converted back to nitrate via a series of oxidative reactions. Many of these oxidations are carried out by autotrophic bacteria which derive energy from the reactions. These organisms appear to be largely responsible for the preponderance of nitrate (70 percent) as the combined form of nitrogen in the oceans.

Specific reaction mechanisms are not known in many cases. For this reason, some of the intermediate steps are shown with an indication of some of the organisms and enzymes known to catalyze the reaction. In other instances, an overall reaction is suggested. For example, plants and many microorganisms convert nitrate into cellular protein, but it is not known whether they produce intermediate products such as nitrous oxide or hydroxylamine in the process. The same is true for reduction of nitrite (NO_2^-) by certain heterotrophic bacteria to ammonium.

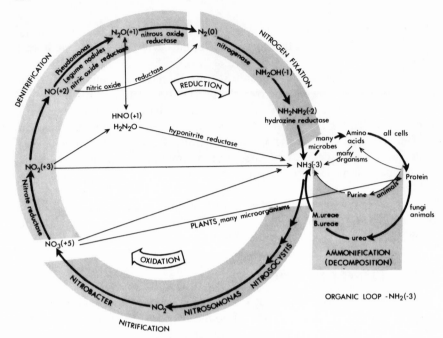

Figure 27. The nitrogen cycle.

Nitrogen gas (N_2) comprises approximately 80 percent of the air in our environment. It is generally unavailable for most biological reactions in its elemental form because of its relative chemical inertness. However, all of the nitrogen that is found in nitrogenous chemicals is presumed to have arisen from gaseous nitrogen, and most of it has been fixed by the action of microorganisms. Many microorganisms including anaerobic bacteria and heterocysts of several blue-green algae have the enzymic capacity for converting atmospheric nitrogen into a biologically utilizable form. The process is called *nitrogen fixation*, and it is much more prevalent in nature than was assumed a few years ago when discussions of the process centered around the symbiotic activities of *Rhizobium* in root nodules of leguminous plants and *Azotobacter* in soils. Nitrogen fixation by root nodules of nonlegumes has been reported as well as fixation by grasses in the absence of bacteria and nodules (Raggio *et al.*, 1959, Bond and Gardner, 1957; Stevenson 1959).

The nitrogen-fixing enzyme system requires an external supply of iron, magnesium, and molybdenum ions. Calcium, boron, and cobalt ions have been implicated as requirements by some organisms under some conditions. Nitrogen gas (N_2) appears to be absorbed by a nitrogenase enzyme molecule which interacts with a hydrogenase enzyme to give a partially reduced nitrogen enzyme complex, as illustrated in (10.1), and is subsequently reduced to ammonium (NH_3) (Mortenson, 1962).

$$H_2 + \text{flavin enzyme} \rightleftharpoons H_2 \cdot \text{flavin} \cdot E$$

$$N_2 + \underset{\text{enzyme}}{M-M} \rightleftharpoons \underset{M-M}{\overset{N=N}{|}}$$

$$\left.\begin{array}{c}\\ \\ \\ \end{array}\right\} \xrightarrow{\text{ATP}} \underset{\underset{M-M}{|}}{\overset{\overset{H}{|}}{N}} - \underset{\underset{\text{pyruvate}}{|}}{\overset{\overset{H}{|}}{N}} \xrightarrow{X} \xrightarrow{Y} \xrightarrow{Z} NH_3 \quad (10.1)$$

In anaerobic bacteria, pyruvic acid is known to be able to supply hydrogen to the reaction. Pyruvic acid is a key metabolic by-product of fermentation reactions which occur in oxygen-deficient zones in the environment. It also is important in the reactions which incorporate inorganic ammonium ion into organic compounds (e.g., amino acids).

The significance of nitrogen fixation is that it affords a means of converting a pristine body of water that is low in both organic materials (oligotrophic water) and in combined inorganic nitrogen to a body of water that is enriched in combined nitrogen. This will then enable organisms, the majority of which are not capable of fixing nitrogen, to become established and promote further colonization of the water. Howard et al. (1970) demonstrated that nitrogen fixation occurs at significant rates in Lake Erie in both the water column and in bottom sediments even though the level of available fixed forms of nitrogen (NO_3^-, NH_4^+) is high enough to support biological activities. Nitrogen fixation in bottom sediments is relatively insensitive to seasonal temperature variations. This indicates that the nitrogen enrichment is a continuously additive occurrence, which could be further stimulated by the presence of carbon fermentation (pyruvate formation) in the sediments as the result of carbon enrichment or organic pollution.

The reverse effects of nitrogen fixation would be the combined processes of *nitrification* and *denitrification* that result in release of nitrogen gas back to the atmosphere from the combined inorganic forms.

10.4. CYCLING OF SULFUR

The overall sulfur cycle is depicted in Figure 28 as a reductive half cycle through which a sulfur atom is reduced from a $+6$ oxidation state (SO_4^{2-}) to a -2 oxidation state (RSH or H_2S) and an oxidative half cycle in which sulfur is oxidized from the -2 oxidation state through elemental sulfur (S^0) back to sulfate.

The reduction of sulfate (SO_4^{2-}) to sulfide (S^{2-}) involves the biological synthesis of several intermediate sulfur-containing compounds and an overall transfer of eight electrons. These reactions can be carried out by any organism which can utilize sulfate salts as nutrients for production of cellular

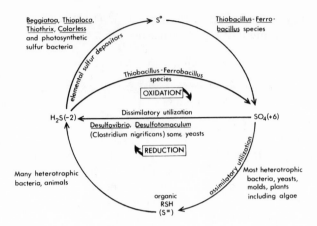

Figure 28. The sulfur cycle.

protein. Many of the intermediate reactions are not well understood biochemically. However, some insight can be gained by examination of the reverse processes, which have been studied in sulfur-oxidizing organisms and are presented in Chapter 11. The initial sulfate reductive reactions have been investigated in species of the dissimilatory sulfate-reducing bacteria *Desulfovibrio* and *Desulfotomaculum*. These reactions involve formation of adenosine phosphosulfate (APS) from adenosine triphosphate (ATP) plus inorganic sulfate ion and the subsequent reduction of APS to adenosine monophosphate (AMP) plus sulfite (SO_3^{2-}):

$$ATP + SO_4^{2-} \xrightleftharpoons{\text{ATP sulfurylase}} APS + PPi \qquad (10.2)$$

$$\text{inorganic pyrophosphatase} \quad \searrow^{H_2O} \searrow 2\,Pi$$

$$APS + 2\,e \xrightleftharpoons[\text{donor not identified}]{\text{APS reductase}} AMP + SO_3^{2-} \qquad (10.3)$$

These heterotrophic organisms ultimately couple the reductive process to the oxidation of a low molecular weight organic acid or alcohol, for example,

$$2\,CH_3\overset{\overset{\displaystyle O}{|}}{C}H—COOH + SO_4^{2-} \rightarrow 2\,CH_3—COOH + 2\,CO_2 + H_2S + 2\,OH^-$$

lactic acid   acetic acid

$$(10.4)$$

The compound APS can be converted by organisms other than dissimilatory sulfate reducers to phosphoadenosine phosphosulfate (PAPS) plus adenosine diphosphate if ATP is available.

$$APS + ATP \xrightarrow{APS \text{ kinase}} PAPS + ADP \tag{10.5}$$

PAPS is the "active sulfate" form which is involved in transfer of sulfate to other cellular constituents such as phenols, carbohydrates, and lipids, for example,

$$PAPS + (\text{lipoic acid}) \; Lip \begin{array}{c} SH \rightarrow Lip \\ / \qquad \qquad \\ \diagdown \\ SH \end{array} \begin{array}{c} S{-}SO_3^- + PAP(H) \\ / \\ \diagdown \\ S{-}H \end{array} \tag{10.6}$$

The dissimilatory sulfate reducers examined do not produce APS kinase and therefore do not synthesize or metabolize PAPS. These organisms therefore must satisfy their assimilatory sulfur needs (e.g., protein syntheses) by reactions other than the PAPS system. See those shown for conversion of hydroxypyruvate to cysteine (10.13). A dissimilatory process implies that the reaction products are not ultimately assimilated into cellular components but rather are deposited as by-products in the environment. In this case, hydrogen sulfide would be produced as the familiar gas having the odor of rotten eggs. If the water is alkaline, the H_2S would react to form a soluble salt, as shown in reactions (10.7) and (10.8):

$$H_2S \rightarrow H^+ + HS^-, \qquad HS^- \rightarrow H^+ + S^{2-} \tag{10.7}$$

$$H^+ + HS^- + Na^+ + OH^- \rightarrow Na^+SH^- + HOH \tag{10.8}$$

If the water contains heavy metal ions, the sulfide would react to form insoluble precipitates:

$$2H^+ + S^{2-} + Fe^{2+} + 2\,OH^- \rightarrow FeS \text{ black ppt.} + 2\,HOH \tag{10.9}$$

Reactions (10.10) to (10.13) are examples which illustrate the microbial conversion of organically bound sulfide (RSH) to the inorganic form (H_2S) after protein is hydrolyzed to sulfur-containing and non-sulfur-containing amino acids.

$$\underset{\text{cysteine}}{HS{-}CH_2{-}CH{-}CH{-}COOH} + H_2O \xrightarrow[\text{desulfhydrase}]{\text{cysteine}}$$

with NH_2 attached above the CH.

$$H_2S + NH_3 + \underset{\text{pyruvate}}{CH_3\overset{\overset{O}{\|}}{C}{-}COOH} \tag{10.10}$$

The above reaction is known to be carried out by *Proteus vulgaris, P. morganii, Escherichia coli, Bacillus subtilis,* and *Propionibacterium pento-saceum,* among others.

$$HS{-}CH_2{-}CH_2{-}\overset{\overset{\displaystyle NH_2}{|}}{CH}{-}COOH + H_2O \rightarrow$$

$$H_2S + NH_3 + CH_2CH_2\overset{\overset{\displaystyle O}{\|}}{C}{-}COOH \quad (10.11)$$

ketobutyric acid

Proteus species

$$CH_3{-}S{-}CH_2{-}CH_2{-}\overset{\overset{\displaystyle NH_2}{|}}{CH}{-}COOH + H_2O \xrightarrow[\substack{\text{demercapto} \\ \text{deaminase}}]{\substack{\text{dethiomethylase} \\ \text{or methionine}}}$$

methionine

$$(10.12)$$

$$CH_3SH + NH_3 + CH_2{-}CH_2{-}\overset{\overset{\displaystyle O}{\|}}{C}{-}COOH$$

methyl mercaptan   ketobutyric

In all of the above reactions, the sulfur has not undergone an oxidation or reduction ($S^{2-} \rightarrow S^{2-}$). It is merely the chemical form that has been altered.

Increased numbers and activity of people involve an increase in the rate of cycling nutrients, and as fixation processes come into balance with respiratory (and combustion) processes the numbers and activities of the organisms bringing about the balance will be stimulated to increase. We can say that increases of autotrophic carbon dioxide fixers such as algae are an inevitable and necessary consequence of carbon dioxide production, and we should interpret algal blooms as a natural sequel to our activities.

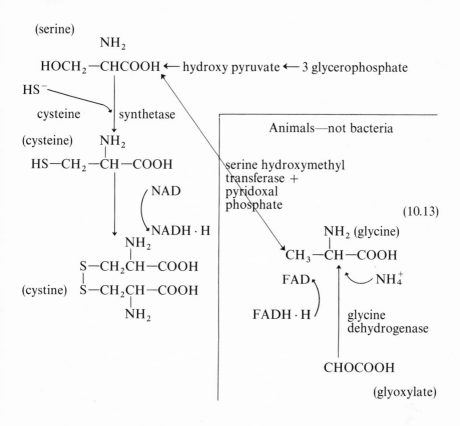

(10.13)

PART III

MAJOR ECOLOGICAL PROBLEMS

Chapter 11

BIOCHEMISTRY OF ACID MINE DRAINAGE

In general, the involvement of biological systems in acid mine drainage can be arbitrarily divided into three categories: (1) the influence of acid drainage on biological systems, (2) the influence of organisms on formation of acid drainage, and, (3) biological means of abatement and treatment. Distinction between two categories of microbes is made on the basis of nutritional requirements. Both types have been discussed extensively in the literature (Dugan and Randles, 1968; Ivanov, 1964; Kelly, 1967; Silverman and Ehrlich, 1964).

Autotrophic microorganisms are those organisms which utilize carbon dioxide, either oxidation of minerals or photosynthesis as their energy, and a few trace minerals and/or vitamins as additional nutrients. This type of microbe, which includes the *Thiobacillus–Ferrobacillus* group of bacteria as well as algae, can therefore grow in a minimal nutritional environment since all of the minimal requirements are readily available in natural water (acid water in the case of *Thiobacillus–Ferrobacillus*).

Heterotrophic microbes are those which depend upon the oxidation of reduced organic compounds for their energy in addition to their cellular carbon requirements. They also have nutritional requirements for minerals and/or vitamins similar to those of the autotrophic organisms: The heterotrophic microbes include the *Desulfovibrio–Desulfotomaculum* group of bacteria that has been recommended as possible agents for acid water treatment (Tuttle *et al.*, 1969). In general, specific differences among species of heterotrophic microbes are reflected in differences among types of organic compounds required by each species. This category of organism is somewhat more fastidious nutritionally than the autotrophic category, and nutritional requirements vary widely.

It should be pointed out that all organisms do not fall neatly into one category or the other. Many organisms are known which have the facility to adapt either to an autotrophic or heterotrophic mode of existence and are referred to as *facultative autotrophs* or *facultative heterotrophs*.

It is generally recognized that acid drainage has a deleterious influence on multicellular species of plants and animals (Riley, 1965). However, many protist organisms (bacteria, algae, yeasts, and filamentous fungi) have the capacity to live in a mine acid environment (Dugan *et al.*, 1970c).

11.1. PRODUCTION OF ACID

11.1.1. The Role of the *Thiobacillus–Ferrobacillus* Group of Bacteria in the Formation of Acid Mine Drainage

The mineral pyrite is often found in close association with coal. When coal is mined, the pyritic material is left behind as a waste product which is exposed to atmospheric moisture and oxygen. Large piles of waste pyrite and low-grade coal often accumulate and are referred to in mining jargon as "gob piles."

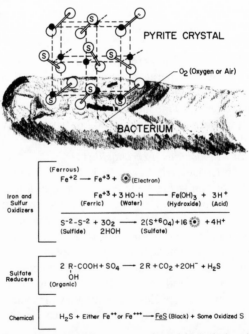

Figure 29. Schematic representation of biological and chemical reactions involved in oxidation of pyrite crystals and subsequent reduction of sulfate.

Chemically, pyrite is a crystal composed of reduced iron and sulfur (FeS_2) and can be illustrated as shown in Figure 29, where the ratio of sulfur atoms to iron is 2 to 1. Electrons shared between the various atoms hold the crystal together. In an oversimplified consideration, the electrons can be extracted from either iron or sulfur and accepted by atmospheric oxygen atoms and the iron or sulfur atoms become oxidized. Removal of a single electron from the crystal, whether from iron or sulfur, destroys the integral properties of a relatively large segment of the pyrite crystal and results in an alteration of the surface area exposure. Ferrous (pyritic) iron is oxidized to ferric iron by losing an electron. The ferric iron thus produced can then associate with water to form ferric hydroxide. The hydrogen ions (H^+) indicated are acid, and their presence is recognized as a lowering of the pH in the stream—a measurement of hydrogen ion concentration. In a manner analogous to that described for iron oxidation, pyritic sulfur (sulfide or S^{-2}) will ultimately be oxidized to sulfate (SO_4^{-2}) by losing eight electrons.

Ferric hydroxide (and there may be several other chemically complex forms of hydroxides and oxides produced) is highly colored and rather insoluble in water, and is recognized as the colored precipitate deposited on the bottom of acid streams.

Bacteria of the acidophilic *Thiobacillus–Ferrobacillus* group can extract and utilize the energy released in the form of electrons from pyrite to partially satisfy their nutritional demands, much in the same way as people extract energy from candy. The bacteria require oxygen and water while they are carrying out the process. The by-products formed as the result of such bacterial action on pyrite are what is recognized as acid mine pollutants, i.e., acid, ferric hydroxides and oxides, and sulfate ions. The acid and high concentrations of some ions in receiving water have a deleterious influence on other forms of life (e.g., fish) which cannot withstand or adapt to the drastic environmental change. The acid and other ions also have a highly corrosive effect on metals which renders the waters in mine areas unsuitable for most industrial and recreational purposes.

Although it is true that pyrite oxidation will take place chemically, it must be pointed out that the oxidations mentioned are accelerated several hundredfold by the catalytic or enzymic action of the *Thiobacillus* or *Ferrobacillus* bacteria. In fact, the only way these acidophilic bacteria can grow and develop is by the utilization of reduced iron and sulfur compounds as nutrients, and the very presence of *Thiobacillus* or *Ferrobacillus* in mine waters indicates that they have already been responsible for converting pyrite to acid. Relative numbers of the bacteria in waters will therefore correlate with amounts of pyrite which must have been oxidized in order to allow that number of bacteria to grow.

Another point to be considered is that the bacteria of concern require H^+ ion and have an optimum activity in a comparatively high acid environment (pH less than 4), whereas purely chemical oxidations of reduced iron and sulfur are drastically retarded below pH 4.

In a continuous flow situation where approximately 10^8 cells per milliliter are continuously being removed, a finite amount of pyritic material is continuously being oxidized by the bacteria. A gross calculation based upon experimental data indicates that 0.16 mole of iron oxidized yields approximately 10^{11} cells, or 0.64 g-moles of iron oxidized will yield the number of cells found in a gallon of water. If the flow of water away from the source is 100 gal/min, then 64 moles of iron would have been oxidized per minute to yield the cells being lost. The efficiency of energy conversion has been reported to be 10 to 30 percent; therefore, $64 \times 3 = 192$ g-moles of iron would be the minimum Fe^{2+} oxidized per minute in the above example; 192 moles of Fe^{2+} is the amount found in about 50 lb of pyrite. These calculations are only intended as illustrations, and no accounting has been made for energy released from the sulfide in pyrite, which is about 10 times greater per mole than that from Fe^{2+} iron. Assuming that all ferrous and sulfide in pyrite were oxidized to ferric and sulfate, it would have required oxidation of 5 to 50 lb of pyrite per minute to yield the acidophilic autotrophs in a stream having a flow of 100 gal/min.

If the above assessment is reasonable, then the basic question to be resolved in this regard is what is the relative significance of microbial pyrite oxidation in comparison to chemical or nonbiological oxidation in a variety of field situations.

If it is concluded that biological oxidation is significant in proportion to nonbiological oxidation, then efforts must be expanded to determine the most effective means of inhibiting the oxidations. One promising means is via specific antimicrobial chemicals, provided that the compounds can be placed in proximity to the target organisms in the field and are relatively specific for the target. Further efforts should be undertaken to study the basic metabolism of the organisms in order to understand what inhibitions are effective, how to use them, and under what circumstances they would be most effective.

In situations where acid formation cannot be prevented, more efficient methods of treatment which appear promising in specific situations are (1) reduction of sulphate to sulfide via anaerobic bacteria, a process which has several "spin-off" advantages that might be favorably compared to chemical neutralization processes, and (2) adsorption of metal ions by biologically produced polyelectrolytes which can then be removed by conventional waste treatment technology.

11.1.2. Mechanism of Action of the Microbiological Production of Acid from Reduced Iron and Sulfur Compounds

Bacterial oxidation of ferrous iron proceeds in the following generalized manner:

$$Fe^{2+} \rightarrow Fe^{3+} + electron \tag{11.1}$$

where the liberated electron is used by the bacterium as an energy source and for the ultimate reduction of carbon dioxide into new cell material.

Ferric iron produced biologically will either react with sulfide non-biologically

$$8\,Fe^{3+} + S^{2-} + 4\,H_2O \rightarrow 8\,Fe^{2+} + SO_4^{2-} + 8\,H^+ \tag{11.2}$$

and result in recycling ferric to ferrous iron, which is then again available to *Ferrobacillus* bacteria as an energy source, or ferric iron will react non-biologically with water

$$Fe^{3+} + 3\,H_2O \rightarrow Fe(OH)_3 + 3\,H^+ \tag{11.3}$$

to form yellow-brown ferric hydroxide precipitates and acid.

Acid is also produced microbiologically as a result of direct oxidation of reduced sulfur compounds. Several metabolic reactions have been elucidated, and these reactions are the subject of an excellent review by Trudinger (1967). In this context it must be recalled that the biological oxidation of sulfide to sulfate involves several intermediary steps, not all of which are completely understood. No single species of organism can be expected to possess all of the enzymes necessary for all known sulfur oxidation pathways. However, the following generalized reactions will serve to illustrate how H^+ and SO_4^{2-} can be produced as the result of metabolism of reduced sulfur compounds by bacteria:

1. Oxidation of sulfur probably proceeds by reduction of sulfur by glutathione (GSH) to glutathione polysulfide (GSS_xH) and subsequent oxidation of the GSS_xH as outlined in (11.4) or (11.5):

$$S^0 + 2\,GSH \rightarrow H_2S + GSSG \tag{11.4}$$

$$S_8^0 + GSH \rightarrow GSS_8H \xrightarrow[O_2 \quad H_2O]{} GSS_6H + S_2O_3^{2-} + 2\,H^+$$

elemental sulfur   thiosulfate

$$\xrightarrow[O_2 \quad H_2O]{} \quad \overset{O_2}{\underset{H_2O}{\bigg|}} \tag{11.5}$$

$$GSS_4H + S_2O_3^{2-} + 2\,H^+$$

$$\downarrow \quad etc.$$

in which a net accumulation of thiosulfate and acid could be produced or further used in biological reactions.

2. Oxidation of thiosulfate (SSO_3^{2-}) proceeds by the overall reaction in 11.6.

$$S_2O_3^{2-} + 2\,O_2 + H_2O \rightarrow 2\,SO_4^{2-} + 2\,H^+ \qquad (11.6)$$

Thiosulfate produced in reaction (11.5) is ultimately oxidized biologically to sulfate and acid through several steps. One likely mechanism is via the splitting of thiosulfate into sulfide (S^{2-}) and sulfite (SO_3^{2-}). Sulfite can be oxidized to sulfate microbiologically in two ways: (a) via sulfite oxidase enzyme (E), which can be generalized as

$$E + SO_3^{2-} \rightarrow ESO_3^- \xrightarrow{\;H_2O\;} E^{2-} + SO_4^{2-} + 2\,H^+ \qquad (11.7)$$

where E represents the enzyme, or (b) through the adenosine phosphosulfate (APS) reductase pathway, which was shown in 10.2 and 10.3 (Peck, 1968).

3. Sulfide ion can be oxidized biologically via reactions similar to those shown in (11.4) and (11.5).

11.2. CONSIDERATION OF MECHANISMS OF PYRITE OXIDATION

Devising and assessing potential controls of acid formation in various types of mines, gob piles, and associated materials is dependent upon an understanding of the processes involved in the oxidation of pyritic materials under the varying and variable conditions existing at the actual sites.

It is appropriate to look at the problem of pyrite oxidation from a somewhat different viewpoint, namely, reaction mechanisms in the oxidation. A minimal reaction scheme [(11.8) to (11.10)] is presented. This allows visualization of the process and the factors that might be important in determining the rates of the process. It serves as a guide in directing attention to various important facets of pyrite oxidation which may not be evident from stoichiometric equations or from reaction kinetics.

$$\hspace{12cm} (11.8)$$

$$
(11.8) \; [2\,H_2SO_3] \left[
\begin{array}{c}
O_2 \rightarrow 2\,H_2O \\
4\,H^+ \\
4\,Fe^{3+} \leftrightarrows 4\,Fe^{2+} \\
4\,e \\
4\,H^+ \\
\longrightarrow 2\,SO_4^{2-} + Fe^{2+} + 2\,H^+ \\
2\,H_2O
\end{array}
\right]
\qquad (11.8 \text{ contd.})
$$

$$
\begin{array}{c}
\tfrac{1}{4}O_2 \rightarrow \tfrac{1}{2}H_2O \\
H^+ \\
e \\
Fe^{2+} \rightarrow Fe^{3+}
\end{array}
\qquad (11.9)
$$

$$
Fe^{3+} + 3\,H_2O \rightarrow Fe(OH)_3 + 3\,H^+ \qquad (11.10)
$$

Some factors are obvious from the stoichiometric equations commonly employed, but other factors are not obvious because of catalytic activity or because they are "balanced out" of the reaction equations. It is also valuable to consider possible reaction pathways because the overall reaction kinetics may or may not reveal the mechanisms, and different parts of the reaction may be kinetically important under different conditions.

It is not the intention of the reaction scheme to describe the mechanism of pyrite oxidation; it is designed to illustrate the probable minimal complexity of the process and provide a basis for discussion and experimental work. Secondly, it is possible that the reaction to $2\,S$ does not involve, as implied by the equation, a removal of $2\,e$ (electrons) from the $2\,S$. The oxidation may involve removal of electrons from iron in the pyrite, the oxidation iron then could pick up electrons from the $2\,S^{2-}$ and be reduced again. This would involve two iron atoms, or one iron atom going through two cycles (also see Figure 18). Thirdly, it is also possible that the reaction from $2\,S$ to $2\,SO_3^{2-}$ takes place in two steps, with the utilization of an oxygen molecule in each, and that these may differ mechanistically:

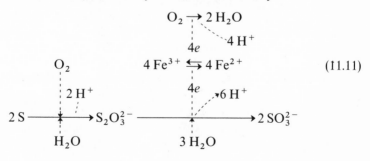

$$
\qquad (11.11)
$$

This would provide an explanation for observed thiosulfate during alkaline oxidation of pyrite, the iron for the second stage being tied up and not available for effective catalysis or electron transfer. It is also possible that this reaction is facilitated by the alkaline environment or by iron complexing so that the normal reaction is switched to this side reaction. Stoichiometric equations of what happens in pyrite oxidation are needed because these provide the necessary base for examining the mechanisms and kinetics of the acid-forming process. These provide certain irreducible minima which must be considered.

Participants in the reaction determine the kinetics of those reactions that are thermodynamically possible. Some of the participants are obvious from the stoichiometrics of the reaction. Some participants are not obvious because they serve catalytic functions or "cancel out" in the overall reaction. Both kinds of participants, and they are not basically different since they both indeed participate in the reaction, may influence the kinetics of the reaction. It is with the second type of participants that knowledge of reaction mechanisms is particularly pertinent and, conversely, where kinetic information can be relevant to determining reaction mechanism.

As an example, if we look at the overall equation

$$FeS_2 + 3.5\,O_2 + H_2O \rightarrow 2\,SO_4^{2-} + Fe^{2+} + 2\,H^+ \qquad (11.12)$$

it is not apparent that ferrous iron has anything to do with the reaction, since the ferrous iron of pyrite is not apparently oxidized. From a kinetic or mechanistic point of view, however, there is growing evidence that the rate-limiting step in pyrite oxidation may indeed be this oxidation of ferrous to ferric iron. This fact is obscured by the rapid reduction of the ferric iron back to the ferrous state again. In these reactions it is relevant to note that the pyrite contains two oxidizable constituents, the iron and the sulfur, and that 93.5 percent (14/15) of the oxygen consumed in the overall reaction is employed in the oxidation of the sulfur component and only 6.5 percent in the iron component, while none of the oxygen consumed in the equation, stoichiometrically speaking, is used in iron oxidation. If iron were, kinetically or mechanistically speaking, the consumer of oxygen, being reduced to ferrous iron, it would have to be oxidized and reduced 14 times during the oxidation of the sulfide portion. This is the extreme. To the extent that oxygen might directly participate in oxidation of the sulfide portion, either by a direct oxygenation or by serving as direct electron acceptor, the number of times the iron portion would be oxidized or reduced would be lessened. To put it another way, seven pairs of electrons must be transferred from the sulfur portion of the molecule to oxygen (two per oxygen atom, four per oxygen molecule). The ferrous iron is capable of transferring only one electron at a time. If it is the only mediator between the oxidation of the sulfur in pyrite and the reduction of oxygen, 14 molecules of ferrous iron would have to be oxidized and 14 of ferric iron reduced for each pyrite molecule oxidized to the ferrous and sulfate levels.

A minimum of two turnovers would be necessary because the maximum amount of oxygen that could be used directly by oxygenation would be $3\,O_2$. It is unlikely that oxidation from the sulfite to sulfate levels involves oxygenation, or even indirect electron transfer to oxygen, so that four more electron transfers through ions are likely. If so, this would limit direct oxygenation or direct electron transfer to oxygen to involvement of $2\,O_2$. In this case, six turnovers of ferrous–ferric ion would be necessary.

To illustrate this, consider the three hypothetical but possible steps

$$(S_2)^{2-} \rightarrow 2\,S + 2\,e$$
$$[2\,H^+ + 2\,e + \tfrac{1}{2}O_2 \rightarrow H_2O] \tag{11.13}$$

$$2\,S + 2\,O_2 \rightarrow 2\,SO_2$$
$$SO_2 + H_2O \rightarrow SO_3^{2-} + 2\,H^+ \tag{11.14}$$

$$2\,SO_3^{2-} + 2\,H_2O \rightarrow 2\,SO_4^{2-} + 4\,H^+ + 4\,e$$
$$4\,H^+ + 4\,e + O_2 \rightarrow 2\,H_2O \tag{11.15}$$

Reaction (11.13) involves going from the oxidation level of the sulfur in pyrite to the level of sulfur itself, which by its nature would necessarily involve electron transfer.

Reaction (11.14) as depicted would be the only step in which direct action of oxygen would be possible, although this is admittedly hypothetical. It could conceivably be an oxygenation, or it could also involve electron transfer either directly to iron or through iron.

Reaction (11.15) does not involve direct oxygenation and probably not direct electron transfer to oxygen.

These reactions have been selected for illustration of what are probably the minimal steps, mechanistically, that can be visualized in oxidation of the sulfur component of pyrite, and the most direct pathway. This is relevant kinetically since it indicates minimal relations of participants in the reaction.

For example, this would suggest the following:

1. Rates of oxidation of Fe^{2+} would need to be two, six, or 14 times the rates of oxidation of pyrite, assuming that this oxidation is rate limiting and that ferric iron acts as oxidant of the sulfur portion of pyrite in these different ways. (See 11.2.)

2. If reaction (11.14) is necessary to the oxidation of the sulfur portion of pyrite, oxidation beyond the level of sulfur could not occur anaerobically

(e.g., with Fe^{3+} as oxidant). Since ferric iron can apparently suffice, in the absence of oxygen, to bring about the oxidation of sulfide sulfur to sulfate, it indicates that oxygenation is not an obligatory step in the oxidation and that all the oxygen found in the sulfate can come from water.

Reaction (11.8) is highly exergonic and essentially irreversible, and hence there are no thermodynamic barriers to its occurrence. Further, it is unlikely that changing concentrations of products can significantly affect the reversibility of this equilibrium reaction with attendant rate effects. That the product may in some other way influence rate is not excluded (e.g., SO_4^{2-}, H^+, Fe^{2+}).

Concentrations of reactants, however, must be carefully considered. It is obvious that this reaction cannot proceed as written unless at least the three reactants are all present, and the influence of at least these three reactants must be considered.

(a) The insolubility of FeS_2 necessitates consideration of an "effective concentration" which would be much less than the total concentration present. There does not seem to be any actual measure of this effective concentration, but surface area would be a closer approximation of "effective concentration" than total concentration, and might be the relevant measure.

(b) Oxygen concentration will be a reaction rate factor provided that one of the other two reactants or both are not limiting, unless some other oxidizing agent is employed (e.g., Fe^{3+}). When it is considered that 3.5 molecules of oxygen are consumed in the stoichiometric reaction, with the concomitant change of pyrite sulfur from a -1 to a $+6$ oxidation state, it is necessary that the reaction be more complex than written.

(c) Water serves both as a reactant and as a reaction medium. Wherever iron is the intermediary in electron transfer between pyrite oxidation and oxygen, water must be the source of the oxygen that ends up in the sulfate. Hence, it is probable that water is more important than indicated by the stoichiometry of the reaction.

All three of these materials have, within limits, been shown to be rate-limiting factors in the kinetics or the mechanism of pyrite oxidation.

Reaction (11.9) is not a highly exergonic reaction; hence, it is readily reversible and therefore allows the reversible oxidation and reduction of iron under the conditions of pyrite oxidation and hence has catalytic potentialities both chemically and biologically. This is not brought out in the stoichiometry of acid formation because the ferric iron measured (or ferrous iron disappearing) is that which enters reaction (11.10). Stoichiometrically, reaction (11.8) may be the only pertinent reaction in acid formation, but, kinetically, reaction (11.9) may be most important.

Reaction (11.10), the hydrolysis of Fe^{3+}, yields the end product of pyritic iron oxidation, and in most circumstances ferric precipitates do not form at the site of pyrite oxidation, indicating that the ferric iron that may be formed by reaction (11.9) at the oxidation site is rapidly reduced again rather than hydrolyzed. Thus, reaction (11.10) can be considered a secondary process.

There seems little doubt that iron oxidation is a significant factor in pyrite oxidation. This may be through reaction (11.8) involving oxidation of the iron in pyrite, which does not show up in the stoichiometric equation, or it may be through iron in solution, as in reaction (11.9), or both. Distinguishing between these two possibilities is important.

In view of the admitted slow rate of ferrous iron oxidation chemically in acidic solution, and the role of this oxidation in pyrite oxidation, the role of bacteria catalyzing ferrous iron oxidation in accelerating pyrite oxidation is readily seen.

The extent to which the ferric–ferrous couple participates in pyrite oxidation via an electron-transferring function will be pertinent to the extent of involvement of iron-oxidizing bacteria in determining rates of oxidation. Minimally, 2 moles of ferrous iron would need to be oxidized per mole of pyrite oxidized, and, maximally, 14 moles would need be oxidized. From our present incomplete knowledge, it would seem that at least 6 moles would be needed.

11.3. BIOLOGICAL MEANS OF TREATMENT AND ABATEMENT

11.3.1. Treatment

A report on the potential use of heterotrophic anaerobic bacteria (*Desulfovibrio* and *Desulfotomaculum*) as a means of reducing sulfate to sulfide in acid mine water has been published (Tuttle *et al.*, 1969). This process has several advantageous aspects: (1) Sulfide will reduce ferric ions to ferrous ions and will precipitate ferrous ions as insoluble FeS, thereby removing virtually all iron from solution. (b) It has also been reported that precipitated FeS is amenable to mechanical separation; which would yield a sludge that could be further processed to yield a sulfide reagent for further use in treating mine water (Zawadzki and Glenn, 1968). (c) Metabolism of the heterotrophic *Desulfovibrio–Desulfotomaculum* group of bacteria also results in a net increase in pH of their environment. (d) Metabolic by-products of the anaerobic bacteria have an inhibitory effect on the chemoautotrophic iron-oxidizing bacteria.

Two primary difficulties must be overcome to accomplish sulfate reduction in acidic waters. Dissimilatory sulfate-reducing bacteria require an oxidation–reduction potential of -150 to -200 mv; therefore, the water must be made anaerobic. Secondly, a source of organic nutrients to supply energy and carbon for the heterotrophic anaerobes is required. The addition

of organic materials is favorable to the establishment of anaerobic heterotrophic microflora in acidic water.

The process of biological sulfate reduction can be manipulated in the
laboratory to increase the overall efficiency, and attempts to scale up the
process seem to be successful. This suggests that such a process could be
developed into a practical abatement method at specific locations. Potential
methods for accomplishing this process are lagooning, design of a facility
similar to those used for anaerobic sewage digestion, and conversion of

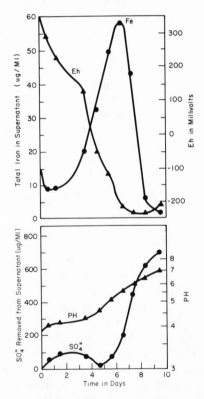

Figure 30. The relationships among pH, Eh, iron concentration (both ferrous and ferric), and the amount of
sulfate reduced in the supernatant of a wood dust culture
containing 400 g of partially degraded wood dust and
2 liters of acid mine water (900 μg of sulfate per milliliter
of water). Prior to incubation, the culture was enriched
with Na-lactate to a final concentration of 0.1 percent
(w/v) and seeded with a 25 ml aliquot of a mixed culture
of sulfate-reducing bacteria. (From *Applied Microbiology*
17:297, 1969.)

certain mines into anaerobic mines where the reduction process would proceed directly in the mine.

The carbonaceous energy source for sulfate reducers could be wood (saw) dust, sewage, waste paper or other domestic waste, algae, and aquatic weeds or other waste vegetable material. Activity of a third group of bacteria is essential to accomplish this process. The third group, referred to as *cellulose digesters*, also proliferates under acidic conditions, and any low-cost cellulose should be able to supply nutrients to the sulfate reducers. The process of microbiological cellulose digestion is exothermic and results in substantial heat liberation, which could be advantageous in treatment of water in cold climates.

The following figures illustrate reactions produced by microorganisms in the highly acid environment of mine drainage. Typical changes of pH, Eh, sulfate, and total iron ions vs. time are shown in Figure 30. These data were taken during growth of a mixed culture that included wood dust and acid mine water which had been enriched with 0.1 percent sodium lactate. The pH increased from 3.6 to 7.0 during a 10-day period. Eh (oxidation–reduction) continually decreased but had a change in rate of decrease at approximately the fourth day. The solution potential became negative between the fifth and tenth days. During the same time interval, the slope of the pH curve increased and an abrupt alteration in the rate of sulfate removal became evident. The concentration of dissolved iron increased markedly for the first 6 days, which can be attributed to increased solubility of ferrous iron as the potential dropped. After 6 days, the Eh approached -200 mv, and the sulfate removal proceeded at a maximum rate during this time period because of precipitation of black FeS (Tuttle *et al.*, 1969). Figure 31 is a diagram summarizing some of the significant biochemical

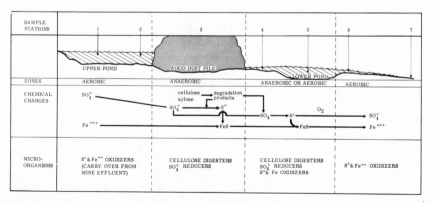

Figure 31. Schematic outline depicting the distribution of microorganisms and the chemical changes they cause in an acid mine water ecosystem. (From *J. Bacteriol.* **97**: 594, 1968.)

reactions found in an acid mine water ecosystem. The net result is a complete cycling of iron and sulfur from the reduced FeS_2 to oxidized $Fe_2(SO_4)_3$ and back to FeS. The FeS can then be reoxidized chemically with atmospheric oxygen. Sample station numerals represent points going downstream from left to right at about 20-yard intervals. The flow diagram shown as Figure 32 illustrates some of the biochemical and chemical interactions in greater detail.

11.3.2. Pretreatment

Processes involving the use of lime or carbonate treatment to precipitate ferric iron or $(FeOH_3)$ from solution are quite sensitive to pH. A high degree of iron removal depends upon a high ratio of ferric to ferrous ions. Autotrophic bacteria (*Thiobacillus–Ferrobacillus*) have been used successfully in

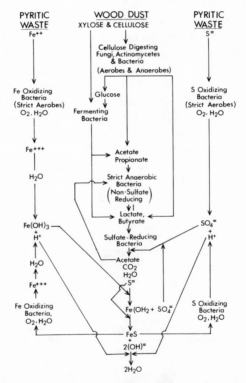

Figure 32. Schematic representation of biochemical and chemical interactions involved in iron and sulfur oxidation and reduction in an acidic mine water ecosystem. (Courtesy of J. Tuttle.)

the laboratory to convert ferrous to ferric ions at pH 2.5 to 3.8 as a prelude to precipitation with limestone. It is anticipated that commercial use of the microbial conversion will require a heat input to raise mine water to optimum temperature for the biological iron oxidation (20 to 25 C). Success of this method of treatment will depend upon availability of lagoons, etc., concentration of iron in solution, and total cost of treatment including the lime precipitation–neutralization and precipitate removal steps.

11.3.3. Abatement

As stated at the beginning of this chapter, it is possible to inhibit metabolism of the autotrophic iron- and sulfur-oxidizing bacteria in the laboratory with the use of chemicals which are quite innocuous to most other living organisms, e.g., alpha-keto acids and carboxylic acids. Preventative methods which utilize antimicrobial agents should prove successful at specific locations. That is, success of prevention of this type pollution would depend upon ability to inhibit causative bacterial metabolism at the origin. Locating and inhibiting the microbial activity should not be a problem in the case of gob piles but may be quite difficult in the case of abandoned drift mines.

The type of antimicrobial agent to be employed in a field situation is a matter that must be further explored. However, preliminary evidence suggests that chemical inhibitors might be practical with reference to cost, availability, and lack of toxicity for organisms other than the iron and sulfur oxidizers (Tuttle *et al.*, 1969).

Chapter 12

POLLUTION AND ACCELERATED EUTROPHICATION OF LAKES

The term *eutrophication* refers to the natural self-enrichment of oligotrophic lakes. The enrichment implies an increase in organic matter in the lakes, which were once extremely low in organic content and also low in fixed inorganic carbon and nitrogen. Presumably, eutrophication is not apropos of increased biological enrichment in the oceans since they were not of glacial origin and did not contain relatively "pure" water. Indeed, the oceans are reported to have had up to 5 percent organic content in their primordial form, a grossly polluted condition by today's standards. At any rate, we have argued that such terms as "pristine," "pollution," and "natural waters" are relative terms that must be considered within the context of people at a given point in time. Freshwater lakes as we know them are increasing in organic content, and the rate at which this is occurring is also increasing as a result of the activities of people.

In lakes where there is (was) insufficient nutrient to support hetero-trophic growth, one would then consider whether there is sufficient fixed inorganic nitrogen to support photosynthesis by higher plants or eukaryotic algae. Under these conditions, there is a tendency for the photoautotrophic bacteria and heterocyst-forming blue-green algae (e.g., *Anabaena*, *Nostoc*, and *Aphanizomenon*) to become established. These organisms fix both CO_2 and N_2 in the presence of light and require only a few dissolved minerals as nutrients for growth (PO_4, CO_4, Mg, Ca, Fe, B, Mo, Mn, Na, Co, Cu, Zn). The blue-green algae are aerobic organisms and produce oxygen while carrying out photosynthesis, although recent evidence suggests that blue-green algae may metabolize heterotrophically in the dark under greatly reduced oxygen tension. The photosynthetic bacteria are anaerobic and grow at depths where oxygen is not available, while utilizing wavelengths of light that were not absorbed by the water and algae above. This initial fixation of inorganic carbon (CO_2, CO_3^{2-}) into an organic form is referred to as *primary production* and the process as *primary productivity*.

Once this type of growth has been established, the cells eventually die and lyse or continue to metabolize and excrete organic by-products into the surrounding water. In either case, the organic material will support growth of heterotrophic organisms and a food chain can develop. Some of the synthesized cellular material is relatively resistant to rapid degradation and tends to settle to the bottom of the lake. Accumulation of sediment on lake bottoms is an aging process which over a period of geologic time will fill the lake. In this sense, all lakes are dying, normally at extremely slow rates, and are destined to become bogs or swamps.

From the above discussion, it is easy to see that the process of heterotrophic growth and sediment accumulation can be accelerated by adding nutrients directly to the lake, bypassing the need for primary production of organic compounds. Pollution then can be considered as a bypass or shunt around primary productivity, and many pollutants can be considered as misplaced nutrients.

Although primary productivity is ordinarily defined in terms of carbon dioxide assimilation, fixation of inorganic nitrogen and sulfur into an organic form can also be considered as primary production. This has particular significance to eutrophication since both nitrogen and sulfur are constituents of enzyme molecules.

Hutchinson (1969), while describing the observation of algal blooms in eutrophic lakes, pointed out that the blooms actually appear during periods of nutrient deficiency in the water. This led to experiments which showed a rapid nutrient turnover between water and sediments. Rigler (1964) demonstrated that ionic phosphorus had a turnover time of approximately 1 min in the epilimnion of a lake in the summer. Hutchinson preferred to consider the eutrophic system of water plus sediment, where the potential for nutrient concentration is high, but may be low in the water at any given time because it is locked up in sediments or bodies of organisms.

It is evident that a eutrophic system will support water blooms of algae and that the frequency of blooms is likely to be related to sediment formation. There is considerable interest in the questions of why blooms suddenly appear and what the controlling factors are. Any of the mineral nutrients listed on page 138 could be growth limiting in a specific lake at any point in time. Also, many environmental variables such as temperature, sunlight, urban and agricultural drainage, and oxidation–reduction potential can influence a sudden and rapid growth of algae.

Many investigators believe that nitrogen and phosphorus are the principal limiting nutrients which govern algal growth in most lakes. This is apparently because agricultural runoff water and treated sewage have a comparatively high content of these two nutrients. Provasoli (1969) has reviewed algal nutrition in relation to eutrophication and pointed out that agricultural runoff and sewage effluents also add sodium, potassium, and other trace metal ions to the water. In addition, water-soluble B vitamins are added to receiving water from this source, and most algae require one of the B vitamins (B_{12}, thiamine, or biotin). Silvey has shown that filamentous blue-green algae grow at the bottom mud–water interface under low dissolved oxygen (< 3 ppm) at 15 C with minimal light (1 lumen or greater). Increased oxygen shifts the metabolism away from nitrogen fixation and alters the cell character. The filaments break loose *en masse* and tend to rise to the surface, giving the appearance of a sudden bloom.

The observation that bacterial blooms preceded *Clathrocystis* blooms in Baisley's Pond, Brooklyn, was recorded for the years 1898 and 1899 by Whipple (1927). This observation has also been made by Dugan *et al.* (1970*b*) in Lake Erie and by Kuentzel (1969), who presented evidence suggestive of carbon dioxide stimulation of growth. The role of bacteria could be one of providing carbon dioxide, vitamins, phosphate, amino acids, organic acids, or other stimulating nutrients, or bacteria could be removing an algal inhibitor or triggering release of a bottom growth of algae. Figure 33 presents some data collected in Lake Erie during the summer of 1969 and illustrates the bacterial production prior to the period of algal bloom in relationship to some chemical parameters in the water. The ratio of phosphorus to nitrogen to carbon in algal cells is reported to average 1 : 16 : 100 on a molecular basis or a ratio of about 1 : 7 : 40 on a weight basis (Megregian, 1967). Based upon this ratio and data of the type presented in Figure 33, Dugan *et al.* (1970*b*) calculated that during a period of algal bloom 16 ppm carbon was tied up in algal cells but only about 5 ppm was missing from the water column. The investigators concluded that a continuous supply of carbon dioxide was entering the system either via anaerobic degradation of sediments or aerobic respiration or diffusion from the atmosphere. Subsequent investigations suggested that diffusion is the least likely source of the carbon dioxide. The bacterial mass present in the lake represented an important factor in availability of nutrients, and the metabolic activity associated with the bacterial mass would result in the release from bound sources of significant quantities of nutrients such as phosphorus, nitrogen, and trace minerals which could stimulate algal growth. It would not be relevant in this case whether the cells were in the sediment, contemporary sediment, or dispersed in the water column. It should also be kept in mind that algal cells never are found in nature in the absence of intimately associated bacteria, which must result in a synergistic or symbiotic influence. Algae are known to photoassimilate 18 to 32 percent of their cell weight from organic compounds such as acetate, glucose, amino acids, urea, casein, and certain other com-

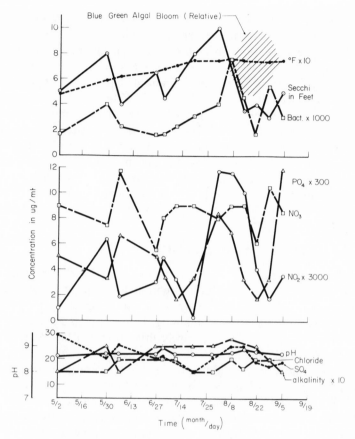

Figure 33. Graph showing number of bacteria, concentration of dissolved mineral ions, and other selected parameters in Lake Erie during the summer of 1969.

pounds which could originate from pollution or from excretions by bacteria and other organisms in the lake. Once an algal bloom has formed, the algae are also capable of excreting substantial amounts of amino acids, peptides, and polysaccharides into the surrounding water. Polysaccharide slime is synthesized by many algae when the ratio of carbon to phosphorus and nitrogen is high (Holm-Hansen, 1968). The significance of polysaccharide to sedimentation is described in the next section.

The significance of algal blooms is in the increased fertility of the water, which may be desirable in some situations and objectionable in others. There is general aesthetic repulsion to the presence of green slime and related possible loss of value on shorefront property. More immediate adverse affects of blooms are production of off tastes and odors in the water, potential clogging of water supply intake filters, the depletion of oxygen in water as the algal cells are decomposed by aerobic bacteria, and the potential toxicity of metabolic by-products produced by some species of blue-green algae (e.g., *Anabaena*). Algal toxin production and its lethal effect on animals has been reviewed by Shilo (1967) and Gorham (1964).

Considerations of algal blooms and their influence on eutrophication must be made with caution since there is a tendency to overgeneralize. Each species of algae has its own unique properties and relationship to the environment, as do all other species of organisms. Experimental data therefore should be interpreted within the context of the experiment, and any single factor may be an overriding influence within that context at that particular point in time.

12.1. LAKE SEDIMENT FORMATION

A wide variety of suspended and colloidal particles enters lake water from the surrounding area. Much attention has been given agricultural land drainage as a source of dissolved nutrients, especially nitrogen and phosphorus because fertilizers containing these nutrients are added extensively to farm land. However, relatively little attention has been given to soil sediment eroded from farm land, or elsewhere, as a source of suspended solids in lakes and streams and its potential in accelerating eutrophication. Biggar and Corey (1969) stated that nutrient losses by erosion tend to be selective in the sense that organic matter and clay particles, which in soil are relatively high in nutrients, are more subject to erosion than are coarser particles. Magnitudes of loss from agricultural land range from 337 lb/acre/ year to 1149 lb/acre/year. Nitrogen and phosphorus losses from soil are both related to organic matter losses. Although nitrate can also leach from soil in a soluble form, phosphate appears to remain bound to soil particles.

Suspended solids in lakes then consist of organic and inorganic particles washed in from surrounding land in addition to organic detritus. All of the suspended microparticles can transport adsorbed nutrients or toxicants as they traverse the water column. Pfister *et al.* (1968) demonstrated that suspended particulate fractions of water are important to microbial relationships in the area of interfaces and biological activity. It is known that particles and molecules collect at interfaces and that enzymatic reactions are concentrated at membranous surfaces. Various suspended particulate fractions were separated on the basis of density, some of which stimulated growth of aquatic microorganisms.

Contemporary sediments, i.e., those which have not yet settled to the bottom, have been collected by Herdendorf (1968) and have a 10 to 15 percent

Figure 34. Electron scan micrograph of contemporary sediment showing diatoms, bacteria, and detritus. (8300 × .)

organic content. The contemporary sediments shown in Figures 34 and 35 appear to be a conglomerate of diatoms, bacterial cells, detritus, and other microparticles. The influence of microorganisms, and the extracellular polymer fibrils they synthesize, on flocculation processes and sediment formation has been examined by Dugan *et al.* (1970*a*). Sediments appear to form as a conglomerate of algal cells, flocculent bacteria, adsorbed organic and inorganic microparticulates, and adsorbed soluble nutrients. Figure 36 illustrates the adsorption of inorganic microparticles on the extracellular polymer fibrils synthesized by floc-forming bacteria. The fibrils also entangle other suspended organisms and adsorb a variety of soluble compounds, both nutrients and toxicants such as chlorinated hydrocarbon pesticides.

Figure 35. Electron micrograph of a carbon replica of contemporary sediment showing
diatoms, bacteria, and detritus. (14,000 X .)

The generalized environmental scheme shown in Figure 37 has been
postulated to explain the interactions among microorganisms, dissolved
and suspended nutrients, sediment formation, and release of nutrients from
sediments. Dissolved organics and minerals as well as microparticles are
carried into the lake from the surrounding environs. The dissolved nutrients
support growth of microorganisms including algae. These in turn synthesize
extracellular polymer fibrils in the form of capsules or slime, which adsorb
microparticles. This association is manifested as floc and contemporary
sediment formation, which settles to the bottom. If the sediments fall in a
shallow zone, they can be resuspended into the water column via wave
turbulence, where they represent a high content of suspended solids and are
capable of repeating the adsorption cycle. If the sediments fall to a deep
nonturbulent zone, decomposition will result in anaerobic conditions, where
hydrolytic and fermentative reactions will result in release of nutrients back
to the water column to stimulate further microbial growth. It should be
emphasized that dissolved organics, minerals, and particulates will continue
to enter the water while sediments decompose. The effect will continue to be
additive as an integral with time, and this will result in a spiraling increase
in the rate of eutrophication. Algal decomposition under aerobic conditions
results in release of 50 percent of the initial bound nitrogen and phosphorus

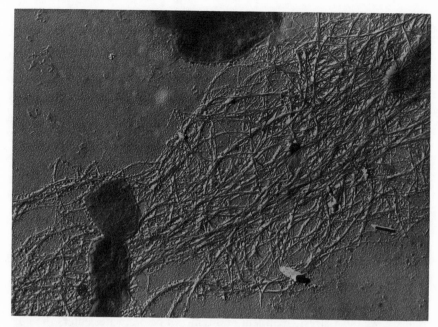

Figure 36. Electron micrograph of the extracellular polymer fibrils synthesized by a gram-negative aquatic pseudomonad showing adsorbed particles of an insoluble inorganic chemical. (15,000 X .)

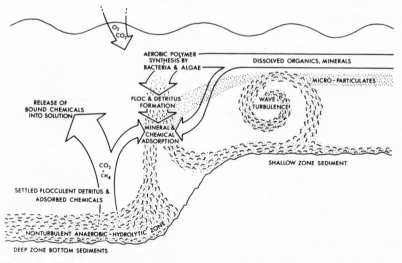

Figure 37. Schematic illustration depicting the interreactions among dissolved organic and inorganic pollutants with suspended microparticles and polymers synthesized by microorganisms. (Taken from Proceedings of the Fifth International Conference on Water Pollution. Pergamon Press.)

within 6 to 12 months (Foree *et al.*, 1970). Corresponding values for anaerobic decomposition are 40 percent nitrogen release and 60 percent phosphorus release. The refractory materials remaining after this period decompose slowly at rates of a few percent per year. Much of the refractory sediment is inorganic and may contain insoluble minerals such as calcium, iron, and aluminum phosphates, depending upon the pH of the water. Bacteria are known to solubilize nutrients and make them available for plant growth, e.g., phosphate from calcium phosphate or apatite (Barber, 1968). This would appear to make attempts to control eutrophication through soluble phosphate removal as a waste treatment procedure somewhat futile except in specified locations where phosphate has clearly been shown to be limiting and where a mineral phosphate reservoir such as apatite is not present.

The fact that we find algae and bacteria together under conditions of accelerating eutrophication probably indicates a symbiotic relationship or at least a passive stimulation among these species as they utilize or produce methane, carbon dioxide, oxygen, organic acids, amino acids, ammonia, etc.

Sedimentation and sediment formation then appear to be key considerations in the eutrophication problem, and these in turn are intimately related to organic and inorganic pollution loading factors. In terms of remedy, we must stress treatment and removal before water enters the lake system, where pollutants become locked into the nutrient cycling regime. To backtrack a step further, we can say that pollutant removal is imperative at the source because there are no such things as nonpolluting treatment additives when used on a large scale. Treatment by additives is merely a system of tradeoffs.

The concept of locked-in nutrients cycling within a lake should alter many oversimplified calculations of the recovery potential for eutrophic lakes. It has been calculated on the basis of flow and retention times that it would take 12.5 yr to flush the pollutants out of Lake Erie if all pollution were stopped immediately. This value will increase significantly when "locked-in" nutrients are considered. The additive effect will also necessitate a reappraisal of the significance of sewage or BOD loading on large bodies of water such as the Great Lakes because the effluent loss is proportionately less than the input.

With regard to "locked in" nutrients, the CH_4 which is released by anaerobic bacteria (see reactions 7.6 through 7.11) appears to be produced in copious amounts in eutrophic water at temperatures considerably below those associated with anaerobic digesters. This implies that strains of methane-producing bacteria are present which function effectively at lake bottom temperatures (5–30°C). There is evidence to suggest that most of the CH_4 produced is oxidized prior to its escape to the atmosphere above the lake surface (see 8.6). Hence the carbon is recycled within the water column and any organic pollutant which can be converted to CH_4 adds to the cycling pool within the lake, resulting in a spiraling increase in the rate of eutrophication.

A similar rational holds for increasing the content of nitrogen within the water column. The rate of fixation of atmospheric N_2 gas is much greater than previously considered. Howard *et al.* (1970) have shown that N_2 is fixed both in the water column and in bottom mud by bacteria and algae and once this is converted to an organic form it is recycled in a lake. Any nitrogenous organic that is added as a pollutant merely adds to the nitrogen pool and accelerates the rate of eutrophication.

REFERENCES

Acid Mine Drainage in Appalachia (1969). Appendix C: The incidence and formation of mine drainage pollution in Appalachia, U.S. Corps of Engineers and U.S. Department of the Interior, 253 pp. Appendix F: Impact of mine drainage on stream ecology, M. Katz, 65 pp.

Albritton, E. C. (ed.) (1952). Standard Values in Nutrition and Metabolism, Wright Air Development Center Technical Report 52–301, 380 pp.

Alper, R., Lundgren, D. G., Marchessauld, R. H., and Cote, W. A. (1963). Properties of poly-β-hydroxybutyrate. 1. General considerations concerning the naturally occurring polymer. *Biopolymers* **1**:545–556.

Asai, T. (1968). *Acetic Acid Bacteria*, University Park Press, Baltimore, 343 pp.

Baptist, J. N., Gholson, R. K., and Coon, M. J. (1963). Hydrocarbon oxidation by a bacterial enzyme system. I. Products of octane oxidation. *Biochim. Biophys. Acta* **69**:40–47.

Barber, D. A. (1968). Microorganisms and the inorganic nutrition of higher plants. *Ann. Rev. Plant Physiol* **19**:71–88.

Berendsen, H. C. (1966). Water structure in biological systems. *Fed. Proc.* **25**:971–975.

Berle, A. A. (1968). What GNP doesn't tells us. *Saturday Review*, August 31, p. 10.

Bernal, J. D. (1966). The structure of water and its biological implications. In *The State and Movement of Water in Living Organisms*, Nineteenth Symposium of the Society for Experimental Biology, Academic Press, New York, pp. 17–32.

Bernal, J. D., and Fowler, R. H. (1933). A theory of water and ionic solution with particular reference to hydrogen and hydroxyl ions. *J. Chem. Phys.* **1**:515–548.

Biggar, J. W., and Corey, R. B. (1969). Agricultural drainage and eutrophication. In *Eutrophication: Causes, Consequences, Correctives*, National Academy of Sciences, Washington, D.C., 661 pp.

Bond, G., and Gardner, I. C. (1957). Nitrogen fixation in non-legume root nodule plants. *Nature* **179**:680–681.

Brown, W. L., Jr. (1958). General adaptation and evolution. *Systematic Zool.* **7**:157–168.

Bryant, M. P., Wolin, E. A., Wolin, M. J., and Wolfe, R. S. (1967). *Methanobacillus omelianski*, a symbiotic association of two species of bacteria. *Arch. Mikrobiol.* **59**:20–31.

Buswell, A. M. and Rodebush, W. H. (1956). Water. *Sci. Am.* **194**:76–89.

Choppin, G. R. (1965). Water, H_2O or $H_{180}O_{90}$? *Chemistry* **38**:6–11.

Cohen, S. S. (1970). Are/were mitochondria and chloroplasts microorganisms? *Am. Sci.* **58**:281–229.

Davis, R. E., Rousseau, D. L., and Board, R. D. (1971). "Polywater": Evidence from electron spectroscopy for chemical analysis (ESCA) of a complex salt mixture. *Science* **171**:167–170.

DeBary, A. (1897). *Die Erscheinung der Symbiose*, Trübner, Strassburg.

Dias, F. F., and Bhat, J. V. (1964). Microbial ecology of activated sludge. I. Dominant bacteria. *Appl. Microbiol.* **12**:412–417.

Donahue, J. (1969). Structure of polywater. *Science* **166**:1000–1002.

Dubos, R., and Kessler, A. (1963). Integrative and disintegrative factors in symbiotic associations. In *Symbiotic Associations*, Thirteenth Symposium of the Society for General Microbiology, Cambridge University Press, London, 356 pp.

Dugan, P. R., and Randles, C. I. (1968). The Microbial Flora of Acid Mine Water and Its Relationship to Formation and Removal of Acid, Water Resources Center, The Ohio State University, Columbus, Ohio, 124 pp.

Dugan, P. R., Pfister, R. M., and Frea, J. I. (1970a). Implications of microbial polymer synthesis in waste treatment and lake eutrophication. Proceedings of the Fifth International Conference on Water Pollution, Pergamon Press, New York.

Dugan, P. R., Pfister, R. M., and Frea, J. I. (1970b). Some microbial–chemical interactions as systems parameters in Lake Erie. In *Systems Analysis for Great Lakes Water Resources*, Proceedings of the Fourth Symposium on Water Resources Research, Ohio State University Water Resources Center, Columbus, Ohio, 135 pp.

Dugan, P. R., Macmillan, C. B., and Pfister, R. M. (1970c). Aerobic heterotrophic bacteria indigenous to pH 2.8 acid mine water: Microscopic examination of acid streamers. *J. Bacteriol.* **101**:973–981. Predominant slime producing bacteria in acid streamers. **101**:982–988.

Dworkin, M., and Foster, J. W. (1956). Studies on *Pseudomonas methanica*. (Sohngen). *J. Bacteriol.* **72**:646–659.

Eisenberg, D., and Kauzmann, W. (1969). *The Structure and Properties of Water*, Oxford University Press, 294 pp.

Evans, W. C. (1963). The microbiological degradation of aromatic compounds. *J. Gen. Microbiol.* **32**:177–184.

Evans, W. C., Smith, B. S. W., Linstead, R. P., and Elvidge, J. A. (1951). Chemistry of the oxidative metabolism of certain aromatic compounds by microorganisms. *Nature* **168**:772–775.

Eyring, H., and Jhon, M. S. (1966). The significant structure theory of water. *Chemistry* **39**:8–13.

Federal Water Pollution Control Administration (1969). Water Pollution Aspects of Urban Runoff, Report No. WP-20-15, U.S. Department of the Interior, 272 pp.

Foree, E. G., Jewell, W. J., and McCarty, P. L. (1970). The extent of nitrogen and phosphorus regeneration from decomposing algae. Proceedings of the Fifth International Conference on Water Pollution Research, Pergamon Press, New York.

Francis, W. (1954). *Coal; Its Formation and Composition*, Edward Arnold, London, 567 pp.

Friedman, B. A., and Dugan, P. R. (1968). Identification of *Zoogloea* species and the relationship to zoogloeal matrix and floc formation. *J. Bacteriol.* **95**:1903–1909.

Gibson, E. T., Wood, J. M., Chapman, P. J., and Dagley, S. (1967). Bacterial degradation of aromatic compounds. *Biotech. Bioeng.* **9**:33–44.

Gorham, P. (1964). Toxic algae. In Jackson, D. F. (ed.), *Algae and Man*, Plenum Press, New York.

Gregory, P. H. (1951). *Proc. Roy. Soc. London* **B138**:202–203.

Harrington, A. A., and Kallio, R. E. (1960). Oxidation of methanol and formaldehyde by *Pseudomonas methanica*. *Can. J. Microbiol.* **6**:1–7.

Hayaishi, O. (1962). *Oxygenases*, Academic Press, New York, 588 pp.

Hayaishi, O. (1966). Crystalline oxygenases and pseudomonads. *Bacteriol. Rev.* **30**:720–731.

Henrici, A. T. (1933). Studies of freshwater bacteria. I. A direct microscopic technique. *J. Bacteriol.* **25**:277–287.

Henry, S. M. (ed.) (1966). *Symbiosis*, Vol. 1, Academic Press, New York, 478 pp.

Herdendorf, C. E. (1968). Sedimentation studies in the south shore reef area of western Lake Erie. Proceedings of the Eleventh Conference on Great Lakes Research, International Association for Great Lakes Research, pp. 188–205.

Hill, D. W., and McCarty, P. L. (1967). Anaerobic degradation of selected chlorinated hydrocarbon pesticides. *J. Water Pollution Control Fed.* **39**:1259–1277.

Holm-Hansen, O. (1968). Ecology, physiology and biochemistry of blue-green algae, *Ann. Rev. Microbiol.* **22**:47–70.

Howard, D. L., Frea, J. I., Pfister, R. M., and Dugan, P. R. (1970). Biological nitrogen fixation in Lake Erie. *Science* **169**:61–62.

Huddleston, R. L., and Allred, R. C. (1963). Microbial oxidation of sulfonated alkylbenzenes. *Develop. Indust. Microbiol.* **4**:24–38.

Humphrey, A. E. (1967). A critical review of hydrocarbon fermentations and their industrial utilization. *Biotech. Bioeng.* **9**:3–24.

Hungate, R. E. (1962). Ecology of bacteria. In *The Bacteria*, Vol. 4, Academic Press, New York, pp. 95–119.

Hutchinson, G. E. (1969). Eutrophication, past and present. In *Eutrophication: Causes, Consequences, Correctives*, National Academy of Sciences, Washington, D.C., pp. 17–26.

Imhoff, K., and Fair, G. M. (1966). *Sewage Treatment*, 2nd ed., Wiley, 338 pp.

Ishikura, T., and Foster, J. W. (1961). Incorporation of molecular oxygen during microbial utilization of olefins. *Nature* **192**:892–893.

Ivanov, M. V. (1964). *Microbiological Processes in the Formation of Sulfur Deposits*, Translated from Russian, 1968 Publ. Israel Program for Scientific Translation for U.S. Department of Agriculture and National Science Foundation, 298 pp.

Johnson, P. A., and Quayle, J. R. (1965). Microbial growth on C_1 compounds. *Biochem. J.* **95**:859–867.

Johnson, W. D., Fuller, F. O., and Scarce, L. E. (1967). Pesticides in the Green Bay Area. Proceedings of the Tenth Conference on Great Lakes Research, International Association for Great Lakes Research, Ann. Arbor, Mich., pp. 363–374.

Joyce, G. H., and Dugan, P. R. (1969). Ester synthesis by *Zoogloea ramigera* 115. *Bacteriol. Proc.*, p. 28.

Joyce, G. H., and Dugan, P. R. (1970). The role of floc forming bacteria in BOD reduction in waste water. *Develop. Indust. Microbiol.* **11**:377–386.

Kallio, R. E., Finnerty, W. R. Wawzonek, S., and Klimstra, P. O. (1963). Mechanisms in the microbial oxidation of alkanes. In Oppenheimer, C. H. (ed.), *Symposium in Marine Microbiology*, Thomas, Springfield, Ill.

Kallman, B. J., and Andrews, A. K. (1963). Reductive dechlorination of DDT to DDD by yeast. *Science* **141**:1050–1051.

Kayushin, L. P. (ed.) (1969). *Water in Biological Systems*, Consultants Bureau, New York, 112 pp.

Kelly, D. P. (1967). Problems of the autotrophic microorganisms. *Sci. Progr.* **55**:35–51.

Kester, A. S., and Foster, J. W. (1963). Diterminal oxidation of long chain alkanes by bacteria. *J. Bacteriol.* **85**:859–869.

Kibley, B. A. (1948). The bacterial oxidation of phenol to beta-ketoadipic acid. *Biochem. J.* **43**:v–vi.

Klotz, I. M. (1962). Water. In Kasha and Pullman (eds.), *Horizons in Biochemistry*, Academic Press, New York, 604 pp.

Kuentzel, L. E. (1969). Bacteria, carbon dioxide and algal blooms. *J. Water Pollution Control Fed.* **41**:1737–1747.

Kuznetsov, S. I., Ivanov, M. V., and Lyalikova, N. (1963). *Introduction to Geological Microbiology*, English edition, C. H. Oppenheimer (ed.), McGraw-Hill, New York.

Laird, A. K. (1969). The dynamics of growth. *Research and Development*, August, pp. 28–31.

Laurent, T. C. (1966a). Solubility of proteins in the presence of polysaccharides. *Fed. Proc.* **25**:1127.

Laurent, T. C. (1966b). In vitro studies on the transport of macromolecules through connective tissue. *Fed. Proc.* **25**:1128–1133.

Lawson, L. R., and Still, C. N. (1957). The biological decomposition of lignin—literature survey. *TAPPI* **40**:56A–80A.

Leadbetter, E. R., and Foster, J. W. (1959). Oxidation products formed from gaseous alkanes by the bacterium *Pseudomonas methanica*. *Arch. Biochem. Biophys.* **82**:491–492.

Leadbetter, E. R., and Foster, J. W. (1960). Bacterial oxidation of gaseous alkanes. *Arch. Microbiol.* **35**:92–104.

Lee, G. F. (1970). Factors Affecting the Transfer of Materials Between Water and Sediments, Literature Review No. 1, University of Wisconsin Water Resources Center, Madison, Wis., 50 pp.

Leshniowsky, W., Dugan, P. R., Pfister, R. M., Frea, J. I., and Randles, C. I. (1970). Aldrin: Removal from lake water by flocculent bacteria. *Science* **169**:993–995.

Ling, G. N. (1966). All or none adsorption by living cells and model protein–water systems: Discussion of the problem of "permease-induction" and determination of secondary and tertiary structures of proteins. *Fed. Proc.* **25**:958–970.

Lippincott, E., Stromberg, R., Grant, W., and Cessac, G. (1969). Polywater. *Science* **164**:1482–1487.

Lukins, H. B., and Foster, J. W. (1963). Methyl ketone metabolism in hydrocarbon-utilizing mycobacteria. *J. Bacteriol.* **85**:1074–1087.

Lundgren, D. G., Alper, R., Schnaitman, C., and Marcjessauld, R. H. (1965). Characterization of poly-β-hydroxybutyrate extracted from different bacteria. *J. Bacteriol.* **89**:245–251.

MacIntosh, R. M. (1971). *A Review of the Recycling Concepts in the Tin-Using Industries*, Tin Research Institute, Columbus, Ohio, 18 pp.

Maxwell, J. C. (1965). Will there be enough water? *Am. Scientist* **53**:97–103.

McCarty, P. L. (1964). Methane fermentation. In Heukelekian, H., and Dondero, N. (eds.), *Principles and Applications of Aquatic Microbiology*, Wiley, New York, 452 pp.

McKenna, E. J., and Kallio, R. E. (1964). Hydrocarbon structure: Its effect on bacterial utilization of alkanes. In Heukelekian, H., and Dondero, N. (eds.), *Principles and Applications of Aquatic Microbiology*, Wiley, New York, 452 pp.

Megregian, S. (1967). More on Lake Erie (Letters). *Environ. Sci. Technol.* **1**:446.

Menzie, C. M. (1969). Metabolism of Pesticides, Special Report No. 127, U.U. Fish and Wildlife Service, U.S. Department of the Interior, 487 pp.

Mortenson, L. E. (1962). Inorganic nitrogen assimilation and ammonia incorporation. In Gunsalua, I. C., and Stanier, R. Y. (eds.), *The Bacteria*, Vol. 3, Academic Press, New York, 718 pp.

Nickerson, W. J. (1969). Decomposition of naturally occurring organic polymers. In *Origin, Distribution, Transport, and Fate of Organic Compounds in Aquatic Environments*, Proceedings of the Fifth Rudolph's Research Conference.

Odum, E. P. (1959). *Fundamentals of Ecology*, 2nd ed., Saunders, Philadelphia, 546 pp.

Okey, R. W., and Bogan, R. H. (1965). Apparent involvement of electronic mechanisms in limiting the microbial metabolism of pesticides. *J. Water Pollution Control Fed.* **37**:692–712.

Ooyama, J., and Foster, J. W. (1965). Bacterial oxidation of cycloparaffinic hydrocarbons. *Antonie VonLeevwenhoek* **31**:45–65.

Orenski, S. W. (1966). Intermicrobial symbiosis. In Henry, S. M. (ed.), *Symbiosis*, Vol. 1, Academic Press, New York, 478 pp.

Pauling, L. (1960). The hydrogen bond (Ch12). In *Nature of the Chemical Bond*, Cornell University Press, Ithaca, N.Y., 644 pp.

Peck, H. D. (1968). Energy coupling mechanisms in chemolithotrophic bacteria. *Ann. Rev. Microbiol.* **22**:489–518.

Pfister, R. M., Dugan, P. R., and Frea, J. I. (1968). Particulate fractions in water and the relationship to aquatic microflora. Proceedings of the Eleventh Conference on Great Lakes Research, International Association for Great Lakes Research, pp. 111–116.

Pfister, R. M., Dugan, P. R., and Frea, J. I. (1969). Microparticulates: Isolation from water and identification of associated chlorinated pesticides. *Science* **166**:878–879.

Pickard, J. P. (1967). Future growth of major U.S. urban regions. *Urban Land*, February, pp. 3–10.

Provasoli, L. (1969). Algal nutrition and eutrophication. In *Eutrophication: Causes, Consequences, Correctives*, National Academy of Sciences, Washington, D.C., 661 pp.

Raggio, N., Raggio, M., and Burris, R. H. (1959). Nitrogen fixation by nodules formed on isolated bean roots. *Biochim. Biophys. Acta* **32**:274–275.

Renn, C. E., Kline, W. A., and Orgel, G. (1964). Destruction of alkyl sulfonates in biological waste treatment by field test. *J. Water Pollution Control. Fed.* **36**:864–879.

Revelle, R. (1963). Water. *Sci. Am.* **209**:92–109.

Revelle, R. (1969). The ocean. *Sci. Am.* **221**:55–65.

Report of the Secretary's Commission on Pesticides and Their Relationship to Environmental Health (1969). Parts I and II, December, U.S. Department of Health, Education and Welfare.

Rigler, F. H. (1964). The phosphorus fractions and the turnover time of inorganic phosphorus in different types of lakes. *Limnol. Oceanogr.* **9**:511–518.

Riley, C. V. (1965). Limnology of acid mine water impoundments. First Symposium on Acid Mine Drainage Research, Mellon Institute, Pittsburgh, Pa.

Robinson, J. R. (1966). Binding of potassium in cells; water transport: Is there a case for active transport? *Fed. Proc.* **25**:1108–1111.

Rousseau, D. L. (1971). "Polywater" and sweat: Similarities between the infrared spectra. *Science* **171**:170–172.

Runnels, L. K. (1966). Ice. *Sci. Am.* **215**:118–126.

Sagan, L. (1967). On the origin of mitosing cells. *J. Theoret. Biol.* **14**:225–274.

Sawyer, C. N., and Ryeknan, D. W. (1958). Synthetic detergents and public water supplies. *J. Am. Water Works Assoc.* **50**:261–276.

Schmidt, D. H., Dugan, P. R., Chorpenning, F. W., and Griffith, P. H. (1970). Antigenic relationships among the floc forming Pseudomonadaceae. *Bacteriol. Proc.*, p. 44.

Shilo, M. (1967). Formation and mode of action of algal toxins. *Bacteriol. Rev.* **31**:180–193.

Silverman, M. P., and Ehrlich, H. L. (1964). Microbial formation and degradation of minerals. *Advan. Appl. Microbiol.* **6**:153–206.

Simko, J. P. Emergy, E. M., and Blank, E. W. (1965). Degradation of linear alkylate sulfonates in sewage. *J. Am. Oil Chemists' Soc.* **42**:627–629.

Sheehan, J. C. (1939). A review of recent studies on the association of water. *Gamma Alpha Rec.* **29**:68.

Snell, F., Shulman, S., Spencer, R., and Moos, C. (1965). Water. In *Biophysical Principles of Structure and Function.* Addison-Wesley, New York, 390 pp.

Sokatch, J. R. (1969). *Bacterial Physiology and Metabolism,* Academic Press, New York, 443 pp.

Sorenson, H. (1962). Decomposition of lignin by soil bacteria and complex formation between autoxidized lignin and organic nitrogen compounds. *J. Gen. Microbiol.* **27**:21–34.

Speakman, J. C. (1966). The fitness of the environment. In *The Molecules,* McGraw-Hill, New York, 158 pp.

Stanier, R. Y., Sleeper, B. P., Motsuchida, V., and MacDonald, D. L. (1950). The bacterial oxidation of aromatic compounds. *J. Bacteriol.* **59**:137–151.

Stanier, R. Y., Doudoroff, M., and Adelberg, E. A. (1963). *The Microbial World.* Prentice-Hall, Englewood Cliffs, N.J., 753 pp.

Stenerson, J. H. U. (1965). DDT-metabolism in resistant and susceptible stable-flies and in bacteria. *Nature* **207**:660–661.

Stevenson, G. (1959). Fixation of nitrogen by non-nodulated seed plants. *Ann. Botany* **23**:622–635.

Stewart, J. E., Kallio, R. E., Stevenson, D. P., Jones, A. C., and Schlissler, D. O. (1959). Bacterial hydrocarbon oxidation. I. Oxidation of *n*-hexadecane by a gram negative coccus. *J. Bacteriol.* **78**:441–448.

Stewart, J. E., Finnerty, W. R., Kallio, R. E., and Stevenson, D. P. (1960). Esters from bacterial oxidation of olefins. *Science* **132**:1254–1255.

Swisher, R. D. (1963). Transient intermediates in the biodegradation of straight chain ABS. *J. Water Pollution Control Fed.* **35**:557–564.

Symposium on Water (1966). Water binding, structure, hydration and solution theory. *Fed. Proc.* **25**:951–1002.

Szent-Gyorgyi, A. (1958). *Bioenergetics,* Academic Press, New York.

Tabak, H., Chambers, C., and Kabler, P. (1959). Bacterial utilization of lignins. *J. Bacteriol.* **78**:469–476.

Thimann, K. (1968). *The Life of Bacteria,* Macmillan, New York, 909 pp.

Trudinger, P. A. (1967). The metabolism of inorganic sulfur compounds by thiobacilli. *Rev. Pure Appl. Chem.* **17**:1–24.

Tuttle, J. H., Dugan, P. R., and Randles, C. I. (1969). Microbial sulfate reduction and its potential utility as a water pollution abatement procedure. *Appl. Microbiol.* **17**:297–302.

U.S. Geological Survey (1970). Mercury in the Environment, Professional Paper No. 713, U.S. Government Printing Office, Washington, D.C., 67 pp.

U.S. Water Resources Council (1968). *The Nation's Water Resources,* Parts 1–7, U.S. Government Printing Office, Washington, D.C.

VanDenburgh, A. S., and Feth, J. H. (1965). Solute erosion and chloride balance in selected river basins of the western conterminous United States. *Water Resources Res.* **1**:537–541.

Webb, S. J. (1965). *Bound Water in Biological Integrity*, Thomas, Springfield, Ill., 187 pp.

Wedemeyer, G. (1966). Dechlorination of DDT by *Aerobacter aerogenes*. *Science* **152**:647.

Wedemeyer, G. (1968). Partial hydrolysis of dieldrin by *Aerobacter aerogenes*. *Appl. Microbiol.* **16**:661–662.

Weibel, S. R. (1969). Urban drainage as a factor in eutrophication. In *Eutrophication: Causes, Consequences, Correctives*, National Academy of Sciences, Washington, D.C., 661 pp.

Whipple, G. C. (1927). *The Microbiology of Drinking Water*, rev. ed., Wiley, London, 586 pp.

Wilber, C. G. (1969). *Metals in Biological Aspects of Water Pollution*, Thomas, Springfield, Ill., pp. 58–72.

Wollman, A. (1962). Water Resources, Publication No. 1000-B, National Academy of Sciences, National Research Council.

Wood, J. M., Kennedy, F. S., and Rosen, C. G. (1968). Synthesis of methyl–mercury compounds on extracts of a methanogenic bacterium. *Nature* **220**:173–174.

Zawadzki, E. A., and Glenn, R. A. (1968). Sulfide Treatment of Acid Mine Drainage, Summary Report on Bituminous Coal Research, Monroeville, Pa., 87 pp.

Zobell, C. E. (1943). The effect of solid surfaces upon bacterial activity. *J. Bacteriol.* **46**:39–56.

Zobell, C. E. (1946). Action of microorganisms on hydrocarbons. *Bacteriol. Rev.* **10**:1–49.

INDEX